COLLIMATING A NEWTONIAN SCIENTIFICALLY

Incorporating the Cave and
Laser Telescope Collimators

PETER R. CLARK - FRAS

authorHOUSE®

AuthorHouse™ UK
1663 Liberty Drive
Bloomington, IN 47403 USA
www.authorhouse.co.uk
Phone: UK TFN: 0800 0148641 (Toll Free inside the UK)
*   UK Local: 02036 956322 (+44 20 3695 6322 from outside the UK)*

Published by AuthorHouse 10/26/2020

ISBN: 978-1-6655-8083-0 (sc)
ISBN: 978-1-6655-8082-3 (e)

Print information available on the last page.

This book is printed on acid-free paper.

CONTENTS

Author's Note.. vii
Preface...ix

Chapter 1   Key Moments and Bold Initiatives.............. 1
Chapter 2   Scaling Up.. 6
Chapter 3   Preface to the Correct, Easier Way
             to Collimate a Newtonian........................... 10
Chapter 4   Preparation of the Telescope 16
Chapter 5   Once More!... 30
Chapter 6   The Development Story to
             Completion Night, 1 October 2014............. 36
Chapter 7   Narrative Version... 41
Chapter 8   The Easier and Definitive Way to
             Collimate a Newtonian................................ 46
Chapter 9   Don't Bottle Out! 51
Chapter 10   Making the Cave Collimator...................... 57
Chapter 11   Suggestions for Matching and
             Upgrading Eyepieces................................... 61
Chapter 12   The New Approach Consolidated.............. 63
Chapter 13   It Just Takes Somebody New 72

Acknowledgements 79
Appendix A   Two Medium-Field Eyepieces.................. 81
Appendix B   A Handy Catadioptric Newtonian 87
Appendix C   Celestial Truant Catching 95
Appendix D   Intuitive Collimation de-Bunked.......... 101

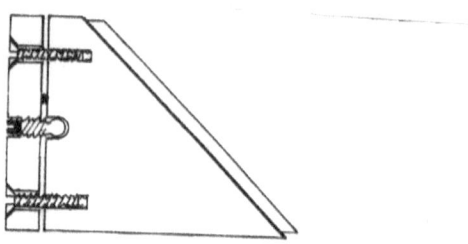

TWO WAYS OF MOUNTING THE SECONDARY MIRROR

The upper one is only used nowadays when the light cone off
the primary mirror can be more parallel. The lower drawing
shows lighter weight possibilities and off-setting built in. Ways
of rotating and moving the mirror in or out are different.

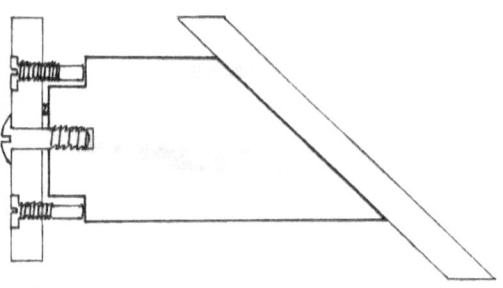

First Edition 2012
Second Edition 01/09/2012
Third Edition 05/03/2014
Fourth Edition 23/03/2015
Fourth Edition Extended 10/03/2017

AUTHOR'S NOTE

The first edition was published as a booklet in 2012. The problems it cured highlighted others for which solutions had to be found. I have taken the opportunities to do this and also to add pages of interest before and after the actual instructions in Chapter 4 that comprise only six pages, including drawings. It is a highly appropriate edition, made so by improved optics creating the possibility of a wider range of targets within one telescope, given the excellent collimation these instructions achieve easily as you progress beyond completion of the adjustments in the first two sentences of stage 1.

PREFACE

Collimation is a relatively new word for the action of aligning mirrors so that light is reflected to where it is wanted. Reflection can be of the sun's rays off a signalling mirror onto a receiving station or potential rescuer, or more to the point, it can occur when light off a dish-shaped telescope's primary mirror is directed towards a second, flat mirror set at 45° to produce a 90° reflection towards a more convenient position for the eye.

The price of a mirror is cheaper than that for an equivalent lens, and the telescope length is reduced to half that of refractors. This century has seen the better optics now available encourage experiments in a further halving of the length of the Newtonian with a variety of lenses affecting price and performance.

The basic Newtonian easily gives the greatest light capture for your money and so should be the most popular, up to a limit of 10″ f/4 for a handy size. Until its secondary mirror became a slice of glass tube bonded to a rod of smaller diameter, as shown in the lower drawing on the inside front cover, the mirrors could rotate sufficiently to need attention by the user.

Another thorn is in trying to collimate the Newtonian secondary mirror or flat mirror in the same way as you do for the in-line Schmidt–Cassegrain, by adjusting any of the usual three adjuster bolts as necessary. Help is at hand, but there is a hindrance. The help is the rotation needed to direct the cone of light towards the eye, which reduces the number of adjusters to one plus a central bolt. Unfortunately, this is countered by few being able to resist the instructions of esteemed mentors to adjust any bolt by trial and error. The lucky and persistent with this common sense can, by compensation for mistakes made unwittingly, achieve collimation with slower telescopes, which is that second- to third-magnitude stars, when magnified, look the same shape as they do when observed by the naked eye. They'll be brighter, of course, but common sense instead of science and engineering applied when there's a fast primary mirror is like flat or earth centralism. It never quite gets the secondary collimated, just passing acceptability maybe, through twisted light paths that come good for a certain temperature or by way of dogged determination.

Chapter 4 is the actual instruction manual. Absolute beginners will first want to apply their telescope's setting up and handling instructions, helped in the final stages by page 26, then use their telescope until seeing something that looks as if it needs fixing. Stages 1–7 scale up gradually from the slowest to the fastest optics for correcting the secondary mirror. In stages 5–7, effects of adjustment will be seen in real time on a star without the retightening delays of the first stages. And we shall see the support for the new approach that really makes things a lot easier as the adjustments become finer. Reliability of the secondary mirror goes from one or two nights to a continuous process. For the primary mirror, adjustment once at the beginning and then once in the middle of a season, at most, is recommended. The classic small Newtonian of f/4.5 or slower can now be all ready for use after completing stage 3 on a well-lit bench, not being waved bye-bye to as the defocussed star is screwed beyond the field of view without any reduction in visible distortion because not all the good materials and knowledge the twentieth century had to offer have been used, including the Frisbee!

Towards the end of 2015, I was getting fed up with forecasts of cloudlessness being repeatedly inadequate for calling members of my astronomy society for a dark site for observing the night sky. By phoning the Meteorological Office, I discovered that although I'd clicked on the information pertinent to my locality at www.metoffice.gov.uk, I was not then going on to the specialised forecasts. These fine-tune the symbols so they show any fine high clouds that put us astronomers off generally, but through a telescope these clouds can actually help with viewing interests such as Venus and bright double stars. Another internet tool you should find well worth opening is the jet stream forecast. From it, one gets a better idea in advance of what magnification is likely to be usable.

Page 27, on the cleaning of mirrors and lenses, is manufacturer and optician supported. I refer to it every occasional time when one or more of them is about to be cleaned. The chapter headed 'Once More!' has its own side elevation drawing that is intended as a window on both standard and more esoteric secondary mirror alignment. I have yet to see a problem that only lateral movement of the secondary mirror will solve.

There is no need to make a big thing of understanding the instructions first. Just use them. Understanding will follow.

CHAPTER 1

KEY MOMENTS AND BOLD INITIATIVES

About 1950, in the eleventh edition of *Norton's Star Atlas*, Arthur P. Norton wrote, 'Sometimes these [secondary mirror adjusters] are three in number, but a better arrangement is a hinged flat mount with a single angle adjusting screw with a knurled head, and an axial screwed pin enabling the flat [mount] to be partly rotated and clamped in position by means of a heavy knurled nut.' He was too polite in rubbishing the fiddling with all three adjusters of the secondary mirror at a time when it may not have been possible to tell the difference, except that with his suggested arrangement, mirror alignment would take less time. Why? Because seven is the number of adjusters on a Newtonian telescope. Seven is the limit of the human brain to make the right empirical choices, and one adjustment affects another slightly. By making the usual two lateral adjusters strict slaves of the central bolt, choices of adjustments become a much more manageable five in total. Norton's one secondary mirror adjuster and central bolt had the potential to end all those blind alleys of 'maybe', 'sometimes', and 'occasionally', and if this had come about, I would not be writing about it.

Sadly, nobody took it up. Then in 1987, there was Walter Scott Houston's famous lament, 'If only amateurs could be persuaded to take collimation seriously.' Any who did so met the wrong methods, which for slower telescopes seemed to work because when focussing in from the symmetrical appearance of a well-defocussed star, no aberrations were revealed amongst the forgiving longer f/6 to f/10 telescopes of the time. Then the bigger mirrors, with the potential of the Dobsonian mount idea, telescope tubes needed to be of practical length in order to offer more than just the one benefit of a cheap and simple mount. Manufacturers went in for between f/5 and f/4, and users found increasing demands on their collimation abilities or were keeping to interests that needed only moderate magnification.

About 1983, Professor Cheshire's eyepiece of 1913 became commercial. The laser collimator appeared in 1995, and my cave collimator appeared in 2009, and we all fell in with the wrong instructions. The laser's manufacturer stayed clear of comprehensive instructions, offering just three lines that were fantastic for helping me damage a mirror edge and a bolt head. In a 2015 review, another was again criticised for acting as if there were no beginners. To be fair, not many makers are wordsmiths, so it may well be 'Over to you, amateur astronomer' with more time.

Between 2003 and 2009, better optics spawned opportunist and established manufacturer attempts to halve the length of the Newtonian telescope, but except for one or two in the easier four-inch class to my knowledge, only the 8″ Cape Newise Newtonian telescope (hereafter referred to as the Wise Newtonian telescope) had the specifications to be successful. Regrettably, 96 out of 100 outsourced optically flat-front glass plates were faulty and caused bankruptcy. My first telescope, a Bresser 4.5″ f/4.4 Pluto, was bought in December 2004.

In 2012, my refusal to blame the telescope when an 8″ f/6 Sky-Watcher was correctly showing the separations of ζ Cancri, a famous triple star, but the Wise wasn't, prompted me into, 'Find a better method quickly, or it's peace for editors.' This threat caused some mysterious force to guide me into screwing the flat secondary mirror out so it was right up against its mounting plate, then off dead equally on all three bolts to complete the collimation mechanically square with a scope for adjustment. Why? Well, apart from angel voices, the machining is always beyond the finest, consisting of fiddling and the blind screwdriving of adjusters 2 and 3 to improve on horizontal alignment. Fine rotation, however, when viewing a slightly defocussed star to make it look as concentric as possible, stage 5, is the correction needed, not imagined horizontal error.

The same applies to fine levelling of EQ-mounted telescopes and to motorised satellite TV dishes when correcting for tracking error. You don't do this, though. Instead you rotate around the zenith overhead, never, ever ruining the unmagnified spirit levelling. Simple does it. It is just the same for correction of a secondary mirror. A desperately needed search for written support from somewhere within the astronomical community spanning 58 years was miraculously successful, the results being referred to in the next chapter.

In 2013, I was using these two improvements. Collimation was seemingly complete, but I was still blaming the seeing too much on double stars. 'Give up on double stars,' one person contributed during my trawl for advice.

In October 2014, the secondary mirror's vertical tilt angle for the f/3 primary mirror had become more critical for my uses of the telescope. Black art deserted me. Panic stations! Unleash the thrice-failed laser collimator to give it the opportunity to redeem itself for this one task only. And it did indeed redeem itself. I was able to see 72 Peg, figure-of-eight split-star trains in M13, and for first time, more impressive than M92, the Moon at full + 1 day with its stunning small craters and impact rays. On 2016 March 30, η Geminorium split for the first time at ×320 - Luis Argüelles's second-most difficult binary for amateurs to split.

Below, I make references to some practical readings that I have found useful. They contain sections for eight-adjustment collimation, any with a nice narrative format drawing one in yet again. The new method achieves six: five adjustments only after the secondary mirror has been positioned. Its elimination of ifs and buts has made the numbered rote style that launches ships' lifeboats successfully into the much easier and more effective format for telescopes:

- David Arditti, *Setting up a Small Observatory*. On page 197 of this work, Arditti shames cleaning perfectionists out of dodging collimation until after the mirrors have been cleaned.
- Vic Menard, *New Perspectives on Newtonian Collimation*. The good drawings on pages 12–14 are helpful for seeing which of the three attitudes illustrated your telescope mirrors have settled into and for assuaging thoughts of precise positioning of optical and mechanical parts.

- 'Testing Optics', by P. J. Violin, on the Cloudy Nights forum. The protocol for the five adjusters now eases his insistence to 'be riveting'. However, his observation 'Joy is not in the well defocussed image for collimating' is even more true today.
- 'Some Collimation Myths and Misunderstandings,' by Nils Olaf Carlin, web.telia.com.

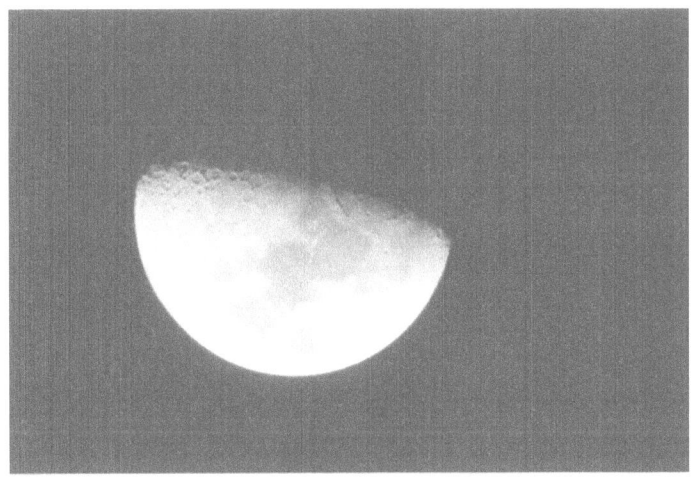

CHAPTER 2

SCALING UP

Instructions for the collimation of Newtonian secondary mirrors have kept everyone in adequate methods that the faster and better optics of this century have called into question. The problem is that they don't scale up to the more exacting demands.

Scaling is a problem common to many things when applying what worked for many years and finding that it does not work when dimensions are changed, when optical ability is improved, and when telescopes are electrically driven. Minimal parts that may be sufficient for a 4" catadioptric Newtonian telescope cost ten times as much as the parts needed to work properly in a congruent 8"—and only if the instructions can scale up. *Catadioptric* describes a telescope that has both mirrors and lenses.

In October 2011, Sky-Watcher 8" f/6 Newtonian telescopes were bought by members of the Blackburn Welfare Astronomy Society, East Yorkshire, for £20 each. I borrowed one for identical specification comparisons with my Cape Newise Newtonian. It was performing better than what I had been achieving so far, in spite of users not knowing about the faulty methods owing to slower primary mirrors, such a tender loving-kindness to practical persons having a determined go. Thus began three years of rediscovery and appraisal of existing methods.

I found it best to begin by inspecting a side elevation drawing, particularly of secondary mirror mounting arrangements, shown on the inside of the front cover. Geometric squareness maintained exactly in the early stages is the first requirement. This means that upon moving the mirror off the mounting plate, for which you can't beat a Workmate bench or an adjustable-height seat, the adjuster bolts must all be slaves that follow the central bolt dead equally. This diligence puts you well on the way to easily achieving the greater accuracy needed later by fast, handier-length Newtonian telescopes.

Also there may well be impressive spin-offs for owners of slower optics reclining smugly! Give it a go. From motorised satellite TV dishes to telescope equatorial mounts, you must not correct by using all three secondary mirror adjuster bolts, meaning that the only attitude adjustment needed is to the vertical tilt bolt in order to correct for 90° not being quite the angle that produces the symmetrical reflection required. This is quite a departure from the usual mantra.

The clear and urgent need of support was found subconsciously on p. 50 of the eleventh edition of *Norton's Star Atlas*, which my mother bought me in 1954 for 17/6p, and in 'Ed's Tracking Travelscope', *Sky & Telescope*, 120/4 (Oct. 2010), p. 70, the only copy of that magazine I've ever ordered—and not knowing what was going to be in it. Both confirmed the need for just one adjuster in line with the focus tube axis. A spring hinge or the pretence of one would be used in place of anything for lateral adjustment. Its adjustment by black art did well until I lost it in September 2014. And so to the thrice-shamed laser collimator a fourth chance was given, in spite of the manufacturer's saying, 'Don't.' It made the vertical tilt adjustment quick, accurate, and reliable.

Independent fiddling about with the two lateral adjusters, usually fitted at 120° to the vertical tilt bolt, ruins the professional squareness achieved by the metal workers, slow optics maybe allowing you to get there sufficiently by persistence eventually, which may be satisfactory for a while. Better by far is rotation whilst viewing a highly magnified star slightly defocussed, for it is nothing more than the delightful fine micrometer or vernier-like solution to the defects revealed by fast primary mirrors that you can now see whilst doing. You will be rotating the secondary mirror about its base or central bolt, identical to the penultimate correction of an equatorial mount, and seeing aberrations reduce, to be removed with a final fine primary mirror adjustment—stages 5–7. Hurray!

Drawings of mirror and tube perimeters in disarray are not needed because with the correct method, anguish and torture may be seen only briefly in the middle of stage 1 through the peephole eyepiece. After achieving symmetry on a single star, you may still be dodging the column if you do not check performance in good seeing of equal-magnitude double stars separated to begin with by about three times your telescope's Dawes's limit. It is also well worth consulting Luis Argüelles's *Finding Your Double Star Limit,* Sky & Telescope Jan.2002 at your own or the atmosphere's pace.

Jupiter's moons are also useful, as is Mars collimated so as not to sublime into three discs when off focus. One has gone through it.

The new approach attitude is helpful to Schmidt–Cassegrain reflector telescopes even though you do adjust any of the three secondary mirror mounting bolts as required. Their f/2 primary and convex secondary mirrors make exacting demands on the beginner sooner, but they're all ready after the complete stage 3. Accuracy with a Newtonian telescope is a more agreeable gradual build-up. I began 'screwing up' the secondary mirror in 2010, but then I thought it useful only for knowing where the mirror would start from. I think photo 003 shows why the Cave Collimator can be so useful in my observatory's chicken, gardening and kennel sharing position.

Before the actual instructions in Chapter 4 is the Preface from November 2014, which covers all Newtonian telescopes and can be helpful for use with other types.

Budget 'scope all well by plumb line.
Some flocking would be good.
A strange setting of the spokes.

CHAPTER 3

PREFACE TO THE CORRECT, EASIER WAY TO COLLIMATE A NEWTONIAN

The instructions for getting it done are in the next chapter. My invention, the cave collimator, is a translucent front cover inscribed with a reticle of centring sights and is used in conjunction with a white 35 mm film peephole canister eyepiece and a doughnut centre mark stuck on to the primary mirror. These objects ensue from Professor Cheshire's 1903 adaptation of the periscope for illumination via the metal peephole eyepiece, and then combine with the cross-wire sights of the sight tube. The origin of the cave collimator is attributed to a black plastic front cover that cracked in 2006, resulting in a new one being made in translucent fibreglass. Then, after three years of limited success with various instructions and a laser collimator, the new cover was accidentally left on, and its use for telescope collimating immediately became clear. The whole telescope soon became an impressive sight tube of almost focal-length accuracy.

The tool is triple purpose, gathering and filtering light to an even tone free from reflections. The cave collimator becomes even more useful to the classic Newtonian telescope before adjustments become finer at faster focal ratios towards the end of star collimating. The convex secondary mirrors of slow catadioptric telescopes have you upping your game with the same joy almost from the start, but with never more than four stages.

The cave collimator works well in ordinary mixed lighting. Even a bed can be suitable for working on. The benefits of using one include greater clarity of the central area by solving the eye-deceiving aspects of the wrong sort of light in which stage 2 may have to be accomplished. And the cave collimator's cross sights give an accurate check. Except for modulated and fast optics, the instructions are quite capable, at the first defocussed star test, of showing the telescope is ready for use. Felt-backed tape is supplied to give a good, smooth fit.

In December 2005, when I was one year into telescopes, I unwittingly took on the demanding optics of the handy 8" Wise Newtonian, my second new telescope with no instructions. 'This must be normal, so get on with it!' Sextant correction was done by way of just one formal method for the same number of reflections, so might telescopes be the same?

The Achilles heel of tools that help with collimation is the instructions. By keeping them to the point, they have required just over four pages plus illustrations and a useful page of tips. The worst reaction is to be intimidated into paying experts to do the Meccano. It is surprisingly easy and keeps the skills alive. Almost forgotten basics has been the key, with no swanning around, along with giving tools a place only where they are useful. Their high priests all need taming somewhat!

Modulated and 'fast' scopes need just three more stages, with which you simply must be chronological. Placing the stages in a circle, starting anywhere, and being lucky can work well enough at slower than classic f/5 ratio, but more adjustments and a lower grade are likely.

As with most things worthwhile, it takes practice, but in ten minutes it is soon achievable. In the instructions, you will see it is recommended that bridges be crossed only when you come to them. Lives are not in danger; damage to mirrors and the receiving of an eyeful of laser beam are both fully prevented by carrying out the new approach to stage 1.

On the first night after you thought you'd gotten it right, you may see just a little more to-do through fast telescopes, but no longer will your fast telescope be collimating night after night and travelling for help. Stages 5–7 will soon have the extra collimation finished.

A white translucent Fuji film canister made of HDPE used as an eyepiece is the best tool for the early stages because it lets light in. It might be all you will ever need. Some instructions for finishing these are on page 15. If one photographic shop doesn't keep any close at hand for art classes, etc., they'll suggest one that does. Other makes, unless you go in for grinding the lips off, are good only as plugs and may not fit well enough. You can also buy them on Amazon.

Some Other Useful Tools

The one- or two-port Cheshire eyepiece is a rather perverse form of periscope that arranges for the target to hit you in the eye. It has cross sights and can be illuminated with a torch, but its smaller field of view makes it less clear. Its best use since 35 mm film canisters is in cooperation with a cave collimator when placed over the corrector plate of a Cassegrain or the objective lens of an expensive folded refractor. Cross sights can then combine to check the total alignment.

The Easy Tester 11 is a metal peephole eyepiece with a highly polished reflective inner surface. It is great for checking if the primary mirror of a slow Newtonian telescope is likely to require fine collimating using a star. The aim with slow telescopes only is to see the doughnut reflecting straight through its central hole out of a now magic-like black surround that is caused by the multiple reflections becoming fainter and fainter.

The laser collimator also is not at its best when used as an early-stage tool. But if there's a problem with setting the vertical tilt angle of the secondary mirror, even the oldest single-port laser collimator becomes the saviour tool, whereas the finest adjuster bolt turn you can make by black art is more likely to set the mirror angle beyond what is needed, etc., with fast primary mirrors.

A red dot finder is good for both visible targets and getting you into the area of those stars invisible to the naked eye. Upon arrival at the area of interest it can be handy to change to an eyepiece giving 1°, 2°, or 3° true field of view and a monocular finder scope of about 3° to 5° for your telescope. The straight-through ones invert the image, whereas the diagonal-viewed ones keep the image upright. So before purchase of any upgrade to basic eyepiece sets, it is simply best to crunch the eyepiece's focal lengths through:

- FSD ×57.3 / telescope fl = true FOV, with field stop diameter being the clear measurement across the back of an eyepiece, easily obtainable from stockists, or see relevant equations on p. 62.

For eyepieces you already have, there's the timing of a star across the field of view. Four minutes of time is 1°. The greatest accuracy is with stars situated within 10° of the Celestial Equator.

Some simple experiments have shown how, with an f/3 primary mirror, just one-eighth of a turn, in on one side and out on the other, of the secondary mirror's two lateral adjusters caused havoc. At f/5, magnified stellar deformity was minimal. Until recently, you could blame the seeing or the telescope, but no longer are there these hiding places, because apart from air pollution, recent rain, and frozen microscopic air particles below about 6000 feet, when they tend to dry out, at least when the sky is clear, the jet stream forecast tells us the seeing to expect before opening the telescope. All you then need are instructions to end the sufferings of the faster Newtonian user. 'A new approach,' 'A better way', and 'Scientific' describe the work that took me from 2009 to break free and progress to completion on 1 October 2014.

Finally, the instructions you may soon be applying now look even easier on paper and only a little less so when hands-on at stage 1; hence the complex drawing shown on page 34 is balanced by the simplest shown in Chapter 13. Thanks to the acceptable state that stages 1–3 will soon get you to, do not be surprised when you find yourself happy to leave fine adjustments using a slightly defocussed star when you don't need to wear gloves or when you've been given a lovely pair of thin mohair runners' winter gloves.

If you have yet to buy your first telescope, then the 4.5" 500 mm focal-length Newtonian, such as Bresser's Pluto and Sky-Watcher's Sky-Hawk 1145P, are very handy.

Inexpensive, they are wide-angle f/4.4 for ease of getting started and seeing all the Pleiades with a field eyepiece 25 mm or a 19 mm wide. The small Newtonian telescope on an upgradable manual equatorial mount is the ideal instrument for the beginner. Lenses turn some into 1000 mm f/8 instruments and may give an upright image for terrestrial use, but they lose some light-gathering ability at night. Deal in new only with specialist astronomy shops!

For ease of setting up and putting away again, do try to avoid the many bolts and the spanner attachment of the cradle rings to the mount that need the 'scope out to get at them. Open the box before purchase! For speed, for carriage, and for preventing the telescope from rolling around during bench collimating, the dovetail type is what is needed. Always offer the 'scope to the mount horizontally with the counterweight set downwards. Go!

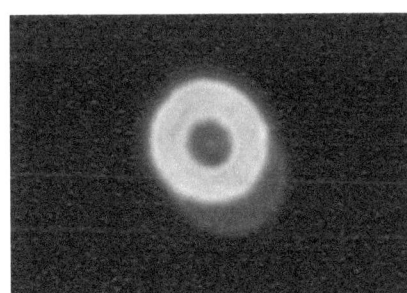

A DSLR one-shot of ι Cancri. The concentric rings, well defocussed, are not as sufficient as they once were. 4″ at. ISO 800. When freshly printed the sixth-magnitude B star should be visible!

CHAPTER 4

PREPARATION OF THE TELESCOPE

Screwdrivers and Allen keys really need to be new and must fit exactly to prevent damage. Bench collimation is carried out mostly with the tube horizontal. Ensure tightness of all assembly bolts and operation of the focuser. A focus tube base which doesn't look square-on should be checked with a plumb line or set square, and correction should be refined further with a spirit level or ruler. Then check it with the peephole cursor and focus tube. Skip this step otherwise, unless it's fun to you, and get straight on into making the optical adjustments known nowadays as collimating, to place the image concentric onto the eyepiece as is.

Making the Peephole Eyepiece

Beginning with a pricker, a 3 mm hole drilled accurately with a new drill bit in the central depression of an all-white Fuji 35 mm film canister bottom is about right, the diameter really depending on your being able to see the edge of the tube through it. If the reflection of the peephole outline isn't too clear, paint its surround red with nail varnish, then inscribe a black cursor line on the inside face with indelible ink. To stop it from dropping down the focus tube, finish by stacking six turns of 12 mm wide holographic sticky metallic fun tape around the canister bottom's outer half and four turns around the top half, but none in the middle.

Primary Mirror Centre Doughnut

For the greatest clarity, it is best to remove any blob-type centre mark with a plastic scraper and an air blower, possibly aided by chilling; your last resort should be a water rinseable solvent on a cotton bud. Paint a paper ring reinforcement with Tipp-Ex, then stick it on centrally. A good method is to cut a hole in the exact centre of a thin paper disk. Cut it to the diameter of the mirror with a circle cutter set to the radius of the doughnut.

The 4–7 Stages for Collimating a Newtonian

It is best to read the instructions for one stage only after completion of another stage. Cross bridges only when you come to them. If you start at stage 2, when nearly aligned anyway, you risk being haunted forever.

Stage 1. Screw the mirrors fully out then back dead *equally* to two turns on primary adjusters. For secondary adjusters, use zero to two turns. Sense the secondary adjusters' contact, and firm them lightly to begin from the point of being *mechanically square* with the telescope for optical adjustments. Do not fall for everything that looks concentric when looking down the focus tube through the peep-hole eyepiece and adjusting secondary mirror adjusters individually just because that intuitive fiddly and regular process looks initially easier than counting equal turns. It may work, but any secondary mirror mounted in a cylinder can 'ding' by over-adjustment, causing it to flake an edge. Making dead-equal turns on the adjusters, experimenting with the centre bolt turns for the best that can be achieved without overlap, and seeing the primary mirror retaining clips are far more important. Allen keys may be the fiddliest, but they are also the most accurate. The accuracy of knobs and screwdriver turns should be increased by marking them in some way. Read no further until stage 1 is completed, etc. Spouting after reading only is not helpful!

Completion of this stage often results in your being able to remove the primary mirror locking bolts if they are a nuisance and you are not shipping the telescope.

Stage 2. Fit the collimating cover and align the peep-hole eyepiece so that the cursor lines up with the focus tube axis. A white wall can substitute, but the centring check of the cross sights at the focal length will not be available. Then, if the doughnut is to one side, *rotate the secondary mirror* to bring it over the focus tube axis, usually by first easing the centre bolt or the first vertical adjuster slightly. There's also the centre spindle method, oval boltholes in the main tube, and the faceplate key, making for glass-mounted secondary mirrors you can't simply grab and turn.

Correction stage 3. Tilt the primary mirror to bring the doughnut over the centre, thereby achieving what is shown in Figure 3 of the ADJUSTMENTS PLANNER page 21. This is about the limit of bench collimating, and by way of this method, slower primary mirrors may be all finished. Star testing now takes you into fine adjustments, if necessary, for the best results.

Stage 4. Using a magnified star. Defocus a second- or third-magnitude star at 120 to 450 times,1 and bring its centre spot central with the effective primary mirror adjuster. The classic parabolic Newtonian telescope is now likely to be fully corrected, or indicating which primary mirror adjuster still needs fine adjustment to improve the defocussed star from depictions 4–8, making it far better for high magnification. The relationship to the adjusters is linear, so it is much more satisfying to work out which is the one to adjust, especially with Allen keys. For this stage, choosing a night when the jet stream is forecast to be clear for your area is best, and when it's been 24 hours since the last spell of prolonged heavy rain. If, whilst you are focussing, the star passes through a *V* shape, comet shape, or funnel shape, repeating stage 1 after taking a breather and returning a portable 'scope to the relaxed situation of a well-lit Workmate bench or master bed after a cup of tea will be the best and quickest solution.

Adjustments planner for Newtonian telescope mirrors

If this shape is more or less horizontal to adjsuters 2 and 3, rotate the secondary mirror until mostly concentric. However, if there is confusion such as by a fan, 'V' or comet shaped aberration, stage (1) will be the best bolt hole to run and start again.

These become more circular as alignment is improved. Any remaining high sports start coming in to line with or opposite one of the adjusters, making final fine correction easier.

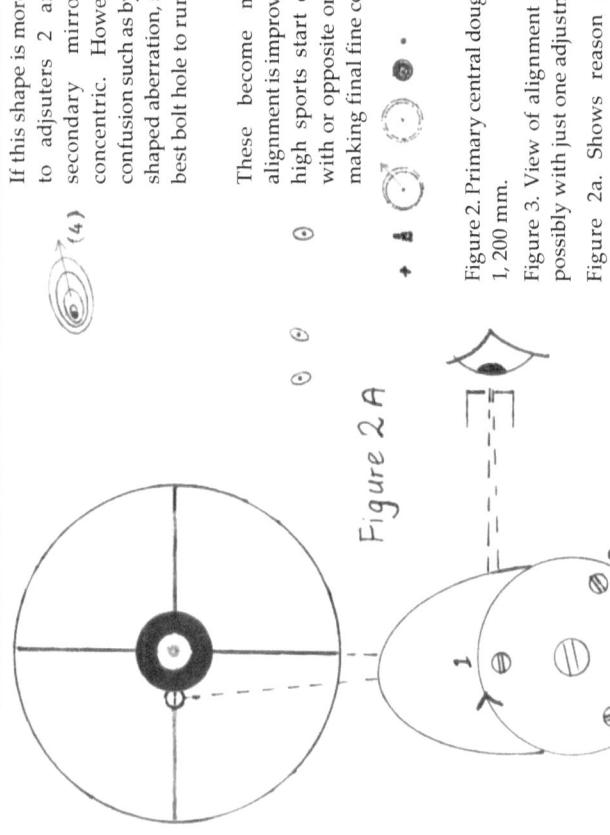

Figure 2 A

Figure 2. Primary central doughnut at f6 and 1, 200 mm.

Figure 3. View of alignment alter rotation, possibly with just one adjustment to finish.

Figure 2a. Shows reason for sideways displacement and arrow showing correction of doughnut position and de-focussed star depiction (7)

Do not shunt with bolts 2 and 3

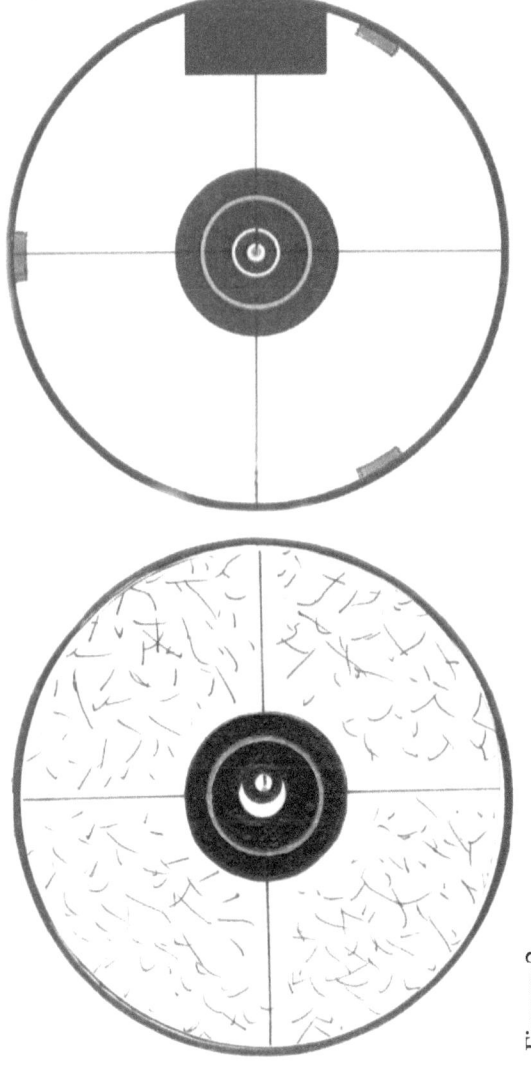

Figure 2

Views through a Peep Hole Eyepiece

Do not be sold by your own finger trouble. That's more likely to occur without a cave collimator. Don't go into buying more tools just yet, if at all.

With f/2 to f/5 imaging Newtonian telescopes and handy spherical mirror types that are modulated to f/6 to f/9, you may well need to correct for the *three remaining errors of the secondary mirror.*

(*NB:* If 'fuzz', or soft blobs, now describe focussed stars, making a simple Hartman mask for analysing will prove spherical aberration clearly. Return to seller. Do the same with a Newtonian that cannot be rotated. There is more risk in not keeping to amateur telescope brands and models advertised in astronomy magazines.)

Stage 5. Check and correct for rotation whilst observing a slightly defocussed star at high magnification. When an adjuster has to be loosened first, use tilt adjuster 1 only, for which you may prefer to level the telescope. Rotation has you seeing the desired results upon re-tightening, as opposed to fiddling with the lateral bolts 2 and 3, which spells disaster for the important mechanical squaring. Too much has to be loosened for the adjusted state to remain steady during the slower blind re-tightening that moves the star.

Stage 6. See depictions 5 and 6. If a pronounced *vertical oval* is seen at any stage, screw the secondary mirror outwards in equal quarter turns on all adjusters until it minimises. Then, for more rounded vertical ovals, try minimal-part turns of the secondary mirror's vertical tilt bolt only to swing the flat mirror towards the focus tube. But with these more critical optics, wonderful black logic trial and error can miss the precise vertical tilt angle needed, time after time. Fortunately, the one task for which a basic laser collimator that hardly moves when you rotate it is most suited is that of tilting its beam vertically only so as to reflect off the horizontal cross-wire sight as imagined by the doughnut whilst viewing through the front. This gets you to within five minutes of being all finished (see page 25). For agreement between tools, and for accuracy with fast mirrors, a twist-lock eyepiece holder/precision centring adapter is strongly advised. Haters of grooved eyepiece barrels will love them too.

Stage 7. Stage 6 will have slightly affected the reflection off the primary mirror, showing a highly magnified star in a degree of oval shape and indicating which one of its adjusters needs turning to round up the image—or for the same aberration but more horizontal, indicating that the secondary mirror should be rotated (depiction 7). Now the star will show which primary mirror adjuster is the way to excellent results on a night when the jet stream is not over your area. Hurray!

Should a depiction like that of Figure 2 remain, leave it. Slight imperfections of any sort that have not been evident with slower instruments suddenly present as offset deviation, throwing the reflections of the laser and the doughnut into varying slight disarray. Believe primarily in the appearance of slightly defocussed stars, and secondarily in Jupiter's moons when available. Then enjoy close double stars until your main interests take over, or the results change you!

Making a telescope that's gone off in use is worse with wrong adjustment. For example, if the doughnut has gone

90° to one side, it is tempting to correct with the primary mirror adjusters. But how can they show displacement in any direction other than in line from an adjuster through to the centre bolt? Only by coincidence and difficulty. The more likely error is slight rotation of the pressure-held secondary mirror.

Back on the bright side of collimation, it is more about aligning the peephole, secondary mirror, and cross hairs on the focus tube axis, but mirrors are not constrained into perfect array. Moreover, the symmetry shown in Figure 3 is only an interim aim and may be final only through f/6 and slower Newtonian telescopes, Cassegrains, and refractors.

Laying the telescope lengthways for car journeys and with the focus tube vertical guards against secondary mirror rotation by gravity. In this regard, it is most helpful for stability if the rings can remain on the telescope at all times.

For telescopes that have to be put away each time, the mounting rings need to be attached to a cradle of the hand-tightened dovetail type that stays on the telescope, not the time-consuming eight bolts and spanners version for fixed observatory and other considerations. These can be a nightmare in storage and transport too. Occasional use of dry lubricant helps with rotation of the telescope tube.

After Successful Star Testing with a 6" Newtonian Telescope and Two Laser Collimators

6″ Newtonian and two laser collimators

If the star test has caused adjustment of the primary mirror to round up a slightly defocussed image as in drawing 8, the temptation to subject your telescope to the instructions of laser collimators or of any aids to correct slight offsets of their positions must be resisted because adjustments indicated by the slightly defocussed star test have taken over. It can even be seen that the doughnut now looks an embarrassment after recently having been essential! Collimation of fast optics can be contradictory!

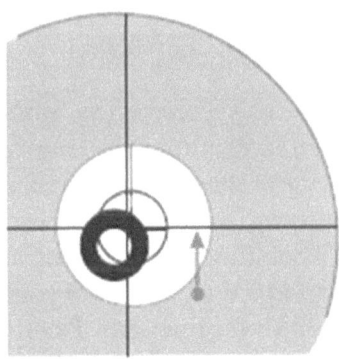

For further improvement, when you see something like this end and peephole combined view through a fast mirror telescope, a laser beam reflection moved onto the horizontal cross hair as imagined by the doughnut, using secondary mirror vertical tilt, only gets you to within five minutes of being all finished.

The secondary mirror reflects black to hide the doughnut until it is within the peephole.

Telescopes needing stages 5–7. The peephole (green) and Cheshire eyepiece (purple) are seen as equals. As shown in stage 7 and depiction 4, you may then be left with a defocussed star shape pointing towards a primary adjuster to turn, or if mostly horizontal, rotate the secondary first. Then with thermal expansion and contraction having become much more linear, you will soon be relaxing with a reliable higher-level collimation with a wider range of seeing than before. Then perfectly concentric and sweet stars seem to live happily alongside slightly misaligned reflections of tools. If this be anguish, it can be assuaged by the view through a second peephole drilled into the side of the central one. Or pay $140 for an autocollimator cap development already so drilled.

Mainly for Beginners to Portable Equatorial Mounted Telescopes

So as to be speedily set up and to be focussed on a star within fifteen minutes every time, paint three white blobs or bed in three damping pads for the tripod feet.

True north alignment may be derived from the Sun's shadow of a vertical stick when at its minimum length, or use the exact meridian passage of time by applying the Equation of Time correction or using a corrected compass. Then, with the counterweights pointing approximately at Kochab and ideally having crafted a 1° true FOV for your telescope with FOV = FSD × 57.3 / focal length, Polaris should become easier to find and work with. A good tool now is a *Planisphere*'s indication of the altitude of any useful star near to the meridian just before you can see it with the naked eye. It has a faint zenith mark from which altitude can be interpolated on the declination scale, or use a builder's or a satellite dish installer's elevation protractor.

Field-stop diameter, the clear measurement across the back of an eyepiece, is easily obtainable from stockists.

NGC 40. An introduction to faint target finding with an EQ mount in manual when observing up into the awkward area close to Polaris. First point the telescope at 50 Cass then rotate in RA until perfectly aligned with β Cass and Polaris. Tiny, fuzzy twelfth-magnitude NGC 40 will be there in any 0.7° to 2° true FOV. Dobsonian users will find it just west of one-third along the straight line from γ Cephi towards γ Cass. A most acute triangle! Triangulation is my favourite way of finding faint targets with a red dot finder. Waiting for this object's culmination in October should be very successful.

Occasional Cleaning of Mirrors, Eyepieces and Corrector Plates

The door to confidence is opened by a documentary in which a segmented VLT is seen being hosed down, so after taking precautions beforehand against dirt, follow these steps for a dropped mirror having a hard landing and, on reassembly, boltholes lining up differently:

1. Remove loose particles with a squeeze-action blower. Then, for more than just dust, rub a ball of cotton wool on cleaned glass. For removal of attached dust alone, strips of low-tack 2″ masking tape may be all that is needed. If not, then—and bearing in mind that air has to have a trickle flow in and out of a telescope—with the

mirror removed, angle it downwards slightly so water doesn't enter via the collars, then pump-spray filtered tap water all over the face. Good illustrations of this stage may be found at www. andysshotglass.com has good illustrations of this stage.

2. Put a couple of drops of smear-less washing-up liquid—'Fairy only', says the window cleaner—in water, preferably filtered, then swab using cotton wool balls, working from the centre outwards in strokes.

3. Pump-spray plenty of distilled water over the face, then dry mostly naturally. Baby cotton buds and unscented paper handkerchiefs may be used to blot only. No wiping now. Do not use toilet paper.

4. For any hardened, rough, clinker-like feel on the glass fronts, use nail polish remover on very soft cotton cloth, wearing thin latex gloves to guard against transfer of hand grease.

Eyepieces

1. As in 1, above, soak a fresh ball of cotton wool in some isopropyl alcohol, then dab, angled slightly downwards, as a means of stopping water entering via threaded collars. Should you attempt more than two eyepieces a day, this great care may desert you.

2. When it looks clean, angle downwards and drench-spray with distilled water.

3. Blot surrounds on paper hankies. Dry naturally. Baby cotton buds and lens wipes can be good to use when moist.

Dew Covers

In still air, the dew covers need to be on corrector plates when dew forms on your car, or when the humidity is over 90 per cent, use a heater band too. At 100 per cent saturated air, all telescopes are best kept closed, if only to reduce the cost of keeping dew off just to see through too much twinkling whilst suffering from dripping water or ice. The freezing level is something you can do nothing about except get up to 6000 feet, where the frozen air particles are drier in settled conditions. So unless the air is drier anyways, less than 90 per cent saturation, an air temperature of 5°C is about the lower limit for magnification to over ×100. Hygrometer and thermometer combinations are very useful. Cheap ones can be bought from nautical instrument suppliers and garden centres.

CHAPTER 5

ONCE MORE!

A chapter that is a sequel to add understanding is in historical order, written as the basis for a talk, and it begins with this:

Stars that still look distorted after an out-of-focus pear shape has been corrected confidently is often the signal for buying another collimation tool.

They've all caused eureka moments, sending their inventors cosmic over one or two useful benefits that can actually corrupt instructions. Awareness that I was still missing something came with triple star ζ Cancri not showing the position angles cleanly. This was the point at which I thought, *What would Galileo and Newton have done first?*

Tools for collimation are mostly best for fitting into existing instructions that are too individualistic, maybe because nobody gets killed by the bad adjustment of telescopes. And yet, may you soon accept that in rushing to rectify, you have largely forgotten the basics. I trawled through everything available, some of it good, for three years, always between perfect sometimes and with a kit of parts to bolt together. The basics were hidden. A lot of hunting was involved!

To refractors first then. Galileo needed only to set the objective lens mechanically square onto its seating. Over about 5″ diameter, quality refractors became mounted on bolts and springs for making fine adjustments. Without instructions to suggest his first move, Newton could only copy refractors. We advance to 2012 and screw both mirrors outwards against their mounting plates, then inwards again by up to a very equal two turns to give scope for adjustments and complete stage 1 basics, and for the ability to rotate the secondary mirror into alignment with the focus tube axis.

With regard to stage 2, rotation into alignment with the focus tube is helped enormously by cross hairs, together with a paper ring doughnut, stuck onto the primary mirror centre, and a peephole eyepiece. This is followed just by centring the doughnut over the cross wires' intersection with the primary mirror's vertical tilt adjuster to complete stage 3 and bench collimating.

The cave collimator and the Cheshire eyepiece are identical in how they can help achieve some of the stages. The Cheshire can be directed towards the best light source, but its reflections are smaller, and eyesight can defocus the true cross wires of the eyepiece. The cave collimator is the best for confused, average, and dazzling light without moving the telescope. By placing it at nearly focal length, its cross sights are usefully independent of the focus tube.

Then all of a sudden you are ready for your star test. A slightly defocussed second- or third-magnitude star using Polaris for AZ/EL mounts, which still looks like depiction 4 or depiction 8, is corrected by adjusting the primary mirror bolt that moves the distortion towards the centre at high magnification and becomes concentric. This places the doughnut off-centre because 90° is said to be slightly over the angle which reflects truly concentric. It also sends the mirrors into some disarray, which matters little nowadays because overlap is prevented by the secondary mirror's being a slice glued on accordingly and positioned to offsetting needs. Any pinching and the defocussed star in the shape of the number of adjuster bolts or the primary mirror's retaining clips pressure is remedied by first easing the adjuster bolts in one-eighth-turn stages up to half a turn equally. When this doesn't work, the tiny bolts of the retaining clips will need easing similarly.

The punishments for column dodgers are that you'll be steering with more than one adjuster bolt and improvement will not be seen before the image goes off the field of view, leading to bad workman judgements. So to be enjoyable, the handier and more complex your telescope, the more you and its manufacturer need basics during a full collimation to either make stage 4 the final one or prevent the three extra stages from adding difficulties.

The drawing for an f/2 Newtonian telescope shows the theory for adjustment of the secondary mirror. As you tilt the primary mirror to make the angle of reflection more acute, you also need to swing the secondary mirror away from the focus tube in sympathy. This adds a touch of tilt automatically. But the tensioned leaf spring or telescopic post mounting for this perfection is delicate and its benefits unproven. Therefore, travel along the centre bolt is almost always completely acceptable.

With regard to Cassegrain telescopes, there is only the secondary mirror to align, using all three adjusters. The first two sentences of stage 1 apply. Then, from the front, decide which one to adjust to make the mirrors concentric with the tube from the best-looking distance for making this judgement. The cave collimator viewed through the focus tube is a very good final check.

Maksutov's secondary mirror is an aluminium surface on the dish corrector plate and cannot usually be collimated by the user because both are very sensitive to tilt error. Any adjusters indicate some relaxation here.

So again, by beginning with Galileo, parabolic Newtonian mirrors can be aligned easily. Fast ones and Maksutovs, Schmidts, Wises, and any other good and handy spherical mirror versions may need a little more fine adjustment, utilising stages 5–7, as seen when brought more into focus by the star's going lopsided, into a vertical oval, clearly saying cheese, or a shape in between. Working in an absolutely chronological way, starting from the basics, is key to keeping secondary mirror adjustments to a minimum. This looks confirmed in the suggestion of *Norton's Star Atlas* from the 1950s that lateral adjusters are not needed. Possibly they are useful for metalworking errors that I have yet to meet in any of the telescopes advertised in the astronomy press.

Finally for any slight deformities in a nearly focussed star magnified 100 to 400 times that you've been tolerating because it's been cold! Feeling like a prisoner to f/3 or to cooling-down time for resembling a slightly elongated pearl sporting a short stalk protruding off the opposite hemisphere? Fear not! The help does not involve a

200-mile car drive. As soon after sunset as is possible in summer is the best time for the quite simple correction sequence of first loosening the centre bolt slightly, then adjusting the tilt slightly and retightening the centre bolt. Now taking your time, inspect a slightly defocussed star, assess it, and react to any difference. Figure 4 can be helpful. This chapter precedes using a laser collimator for this purpose only.

The secondary mirror always gets tilted the wrong way first, but by now most things collimating will be in your stride, including mountings that need the tilt adjuster to be loosened before the centre bolt can be. Then there are those that, when loosened, allow rotation when it is not wanted. They'll have some other good redeeming feature like oval boltholes that can make rotation easier.

Hurdles overcome? All the very best then for you and your telescopes.

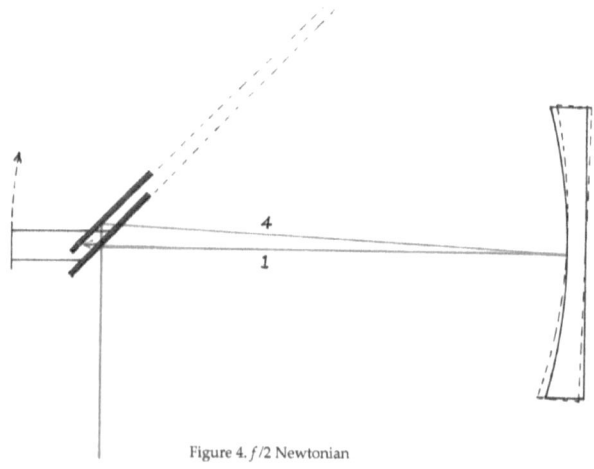

Figure 4. *f*/2 Newtonian

Figure 4 **f/2 Newtonian**

M5 5 Serps. STF1930 overexposed. DSLR 30 secs

CHAPTER 6

THE DEVELOPMENT STORY TO COMPLETION NIGHT, 1 OCTOBER 2014

It is the tenth year since the word *collimation* first entered my vocabulary. Steps old and new are in, and some are out.

For instructions to be accurate, reliable, and quick, they must be practical and to the point without maybes. Tools or aids are there to be incorporated into good existing instructions for the telescope. The trouble is putting what has been done well onto paper. For my laser collimator, there were just three lines on the subject after more than two pages of how to get it ready.

There are some extra stages that begin to be needed from primary mirrors faster than f/5. When you focus in at high magnification, the sweet symmetry can change into an oval shape, and when you are fully focussed, a black spot and slight birdies appear. 'You can never get them quite collimated'; Give up on double stars'; 'Get a refractor. That should do'; 'It's faulty; send it back,' is read. Four of these suggestions are concerned with slower primary mirrors.

Nevertheless, for the beginning of the previous four seasons, I'd been able to apply the right amount of vertical tilt to the secondary mirror, but late this summer the ability deserted me. What is hobbling me? Why have I lost it? A handy impression I did get was that in good seeing, that is the jet stream not forecast to be overhead, and no freezing air at low level, a hexagonal mask for telescopes in plywood or cardboard does very well indeed by redirecting spikes and reducing A star brightness of difficult binaries, as opposed to close ones of equal magnitude.

But when a dew cap is needed, that becomes a hassle. So to resume getting the job done properly again somehow, please, we need a solution. Two years ago, deciding again to follow my basic laser collimator's vague instructions for the secondary mirror, I discovered that it did not work this time either, although unwittingly I must have backtracked to maybe one-thirty-second of a turn off the start position, because by some mysterious black art, the primary mirror needed just a slight adjustment (depiction 8). A small mercy, but not a good basis for instructions! What if finer vertical tilt adjustment could somehow be achieved? Fast primary mirrors are very sensitive to vertical tilt of the secondary mirror, and if a slightly defocussed star magnified to at least half the diameter in millimetres of the primary mirror diameter keeps showing the ability to hit the precise angle required has deserted you, what do you do next?

So to the laser collimator a fourth chance was given. Tilting the secondary mirror absolutely vertically to reflect the beam off the horizontal cross hair as imagined by the doughnut was the answer. Then came rotation, using a slightly defocussed star, to achieve the least oval shape, followed by turning the indicated primary adjuster to bring the image into sweet and concentric focus, all the time under high magnification, having granted the laser collimator a place at last! A twist-lock eyepiece holder is strongly recommended for the repeatability it offers. Haters of grooved eyepiece barrels love them for saving the toes as well as the eyepiece itself.

A slight optical centre deviation from the geometric that hasn't been an evident 'slower downer' suddenly presents upon changing to telescopes with fast primary mirrors. Their extra corrections throw the reflections of the doughnut and laser collimator into some disarray. Again, once the laser has played its part, rotation on a slightly defocussed star may be needed until any horizontal-looking oval almost saying cheese becomes as concentric as possible. Then turn the indicated primary mirror adjuster to bring the image circular and sweet *with any oval off-centred shape's direction now being helpfully linear with one adjuster or a pair of the adjusters on the backplate.*

As soon as you even slightly loosen a spider-mounted secondary mirror, it begins wanting to do its own thing, catlike, and completely unlike the more reliable glass-mounted ones. These offer ease of manoeuvre and retention par excellence. Faith by extra printed words is not needed; they are set up for doing the collimation just right the first time.

Collimation of a laser collimator needs a white sheet of cardboard placed about 6 feet from something such as a vice that you can use for cutting a notch in each cheek so as to take any flange, a piece of wood cut to just under the width of the collimator, and a pencil. The instructions for this that came with my telescope were corny, but I suppose if you have no choice, they really can be wedged in between the *V* of an opened book and squeezed between a lot of other books in a bookcase. I just happen to already have a Record quick vice fixed to the workbench in the right position. A similar cradle can be made from wood.

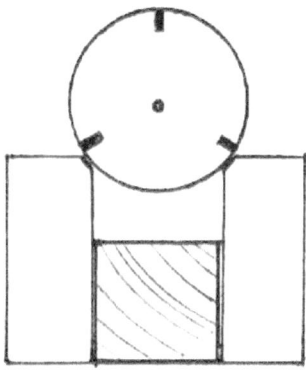

The better a Newtonian design is in terms of handiness and correction, the more the cave collimator benefits will extend to beyond providing a white background as their secondary mirrors tend to be mounted in glass. Maksutovs, Schmidts, Wises, and some Vixen optics really can use the near focal length sight lines set at the most accurate place for them. For the open-fronted Newtonian telescope during a full collimation, there is the tricky step of gingerly grasping and rotating the secondary mirror to relax into before accepting the gallows for the first time. Keeping to no more than one-eighth of a turn loose on the adjusters helps you retain control.

For anyone preferring hearing verbal instructions to following written instructions, keep one hand and an eye at the focus tube and the other on the screwdriver.

With a mostly accurate but slightly fiddly Allen key, or a Bob's Knob, only the person at the eyepiece can see what is being talked about. A flip mirror has possibilities, but the best demonstration way is to read aloud rote style to the pupil hands-on at the eyepiece. A reader who knows nothing about the bad old days is possibly best until it all simply becomes lines easily remembered forever and offering great rewards.

ι Cancri 4.1, 6.0 31" DSLR one-shot

CHAPTER 7

NARRATIVE VERSION

Although in a format that can make for interesting and laid-back reading, collimation is really best achieved through obeying the numbered rote style of instructions. You soon learn by use and become able to pass the instructions on clearly.

I begin this narrative version with the objective lens of the refractor set mechanically square onto its seating by screwing a retaining ring down, and that's it until over about 5″ diameter, when high-quality ones may well be mounted on adjuster bolts as with the secondary mirrors of Newtonian and Cassegrain optics. We need to screw both secondary mirrors outwards onto their mounting plates to bring them into mechanically square alignment given normal decent engineering. Then from all adjusters just touching the plate, follow the centre bolt inwards dead equally this time to two turns off, or as the manufacturer suggests. Wrong methods, lack of awareness, and modern tools try to miss out on these imagined horrors that are just follow-ons of today's sadly missed nuts-and-bolts assembly of making your own telescope.

Beginners, a group which in 2006 I was included amongst, upon successfully collimating a new laser collimator can then cause damage easily by using their untried instructions for aligning the mirrors, which instructions are often disarmingly brief. For Cassegrain users, there is now a very expensive bolthole tool to run to, unnecessarily in my view, whereas with a Newtonian telescope, thanks to rotation around the centre bolt, the secondary mirror needs only one adjuster for vertical tilt, but you never see the arrangement. I found it in 'Ed's Tracking Travelscope', *Sky & Telescope*, Oct. 2010, and *Norton's Star Atlas* (11th - 15th edn, *c.*1950–66), p. 50, with angels and my own timely realisation beforehand.

Collimation of a Cassegrain's secondary mirror soon has you into fine adjustment of any of the three mounting bolts that you decide carefully will make the mirrors' and the tubes' perimeters appear concentric when viewed from an ideal distance in front of the telescope in good outdoor lighting. When looking down the focus tube of a Newtonian telescope through a peephole, because of the different shapes and sizes and positioning decisions that go into gingerly sticking the flat secondary mirror onto a rod, you only need to make everything look as concentric as reasonably possible with nothing overlapping and to see the retaining clips of the primary mirror. Strict equal turns of all three secondary mirror adjusters achieve this as they follow the centre bolt. The two lateral adjusters are never more than exact light-touch followers, which makes for no image shift as you go in and out of focus later on a star. The drawings for two secondary mirror arrangements on the inside front cover show how to begin stage 1 for two types of mounting.

Now see if the secondary mirror needs to be rotated so that the peephole cursor and doughnut reflections coincide on the focus tube axis, then centralise with primary mirror vertical tilt adjustment and so to the star test. With a Newtonian telescope slower than f/4, this test will, so long as the screwing has not been dodged, seldom indicate more than the need to slightly adjust the one primary mirror bolt to achieve fully concentric and sweet alignment. Faster Newtonian primary mirrors may need up to three additional stages. For seven years I felt I had been able to hit the exact amount of secondary mirror vertical tilt needed, then by September 2014 I'd lost it! The beam of a resurrected laser collimator was soon set to reflect off the horizontal axis as imagined by the doughnut. This was entirely against the manufacturer's advice, which I had been obeying from early in 2006.

Whatever other uses the lateral secondary mirror bolts may be fitted for, rotation at high magnification of a star of magnitude 2 to 3 does it far better. And you can see how it is going to look as you tighten just the centre bolt, then whether distortion within 45° off-axis still remains to be taken out with a final small primary mirror adjustment. These are really nothing more than fine repeats of bench stages 2 and 3, continuing not to be difficult through to the finish on a motorised EQ mount, or for a Dobsonian by using the Pole Star. When the secondary mirror is mounted in a glass front, rotation is always under easy control.

You are pretty well finished when you don't have to make any adjustment at the start of the next night, and you're totally finished when bleeder current spikes during temperature settling are so circumferential as to be almost mythical and when distortions don't occur when the telescope is moved around the sky. This better method seems to contribute greatly to expansion and contraction, these having become more linear, and is easier and much more reliable.

Newtonian secondary mirrors are off balance by nature and, rather like pendulums, prefer to travel with the telescope lengthways and the focus tube vertical. I think this says rotation is the mostly likely correction after casual stowage than a journey that renders doubtful the benefit of travelling to pay to have it put right. The Newtonian telescope's focus tube can always be arranged to be in a handy position for viewing. Half-length modulated ones on shorter mounts will be steadier, and with collimation solved, good ones must make a comeback from their rushed entry into the new millennium. Dobsonian mounting is another way of solving problems with size without using steps, but equatorial is a prerequisite for imaging.

Here are two analogies to conclude this chapter, the reading of which can be left until you are well up and running. A secondary mirror standing on its end can be visualised as being identical to a motorised satellite TV dish or a telescope's EQ mount with their bases pointing at the zenith. Once they have been levelled by protractors, spirit levels, and plumb lines, further fine levelling is not achieved by yet more attempts. The solution to depictions 5–7 on page 21 is rotation around the centre bolt to round them up as much as possible. The same is achieved by radio signal peaking when a satellite dish is swung in azimuth onto an east or west transponder to correct the tracking in right ascension. Telescopes have the sufficiently defocussed star for seeing it go symmetrical as you rotate the secondary mirror to get within reach of the final small primary mirror adjustment. I say 'sufficiently' because you reach a point in defocussing at which any fine distortion becomes obscured. This is said by manufacturers to be good enough. It can be adequate with slow telescope optics. They never seem to mention collimation at finer defocus.

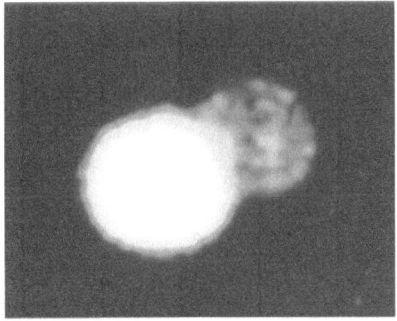

ι Cancri. The more likeable result of overexposure, enlarging in Picasso, then removing chromatic aberration from the A star in Photoshop.

CHAPTER 8

THE EASIER AND DEFINITIVE WAY TO COLLIMATE A NEWTONIAN

In December 2005, after a year with a 4.5″ Bresser Pluto to test my interest, I took on unwittingly the demanding optics of the handy half-length 8″ Wise Newtonian telescope because its confident advertising style mirrored my own confidence throughout the 1970s for a loudspeaker design. It was time to taste my own medicine! 'Eight inches' was the answer given at an East Riding Astronomers meeting in Beverley to the question of what the minimum size was for doing useful things as an observer. Having been a sailor makes me well up on sextant optics, and always quickly on when necessary to inspire opticians into getting my varifocal lenses right without extra cost.

The trouble with collimation of the Newtonian secondary mirror in the beginning is part of the assembly, and therefore this is overlooked by everyone, including twice by me, so the start of the given stage 1 is invariably wrong! With Cassegrain collimation identical at the start, this is amazing. Don't they talk to each other? And coming up fast is that, other than a centre bolt, the Newtonian secondary mirror has been found to need only one adjuster. This realisation emerged out of my subconscious in March 2012 during some difficulty, then was seen later in *Norton's Star Atlas and Telescopic Handbook* of 1950–66, 11th–15th editions, and 'Ed's Tracking Travelscope', *Sky & Telescope*, Oct. 2010, p. 70, the only edition of that magazine I've ever bought! The slower Newtonian primary mirrors are tender loving-kindness up to nearly 2 per cent out. But when the tube length is halved for handiness or wide angle imaging and maybe general-purpose f/5 to f/7 regained with lenses, collimation really becomes a lot finer. Then, paradoxically, the defects become easier to see during star testing. On the bench there will still have been the need for some easing in towards the fiddly grasping bit with the secondary mirror, relaxing it before mounting the gallows. So upon having accomplished all of stage 1, you are very much entitled to celebrate, having become mighty in the trickiest part of collimation and having propelled yourself into a new comfort zone!

There are tools for assisting with collimation. Inventor euphoria has to be reined in. Expect just a slot or two of great usefulness from them! The peephole eyepiece and any spider wires have been doing their part for many years, as has the sight tube with its own cross sights. For when light is a problem, causing a telescope to be moved inconveniently, there are now three solutions. In 1913, only a year after the periscope was invented, Professor Cheshire saw the target could hit you in the eye up the periscope via the mirrors, so he adapted it for helping to correct telescopes. Ron Arbour's company began manufacturing in the 1980s. The laser collimator first appeared in 1995. In 2006, I had to dope a new cap over the telescope in fibreglass because I'd failed to ask if anyone had a lathe I could use! It then took another three years for me to forget to remove it for collimating. Suddenly, amidst confused workshop lighting, indoor collimation became a joy. Cross sights were added, and with the laser's doughnut thank you and a peephole eyepiece, I turned them into the clearest and most accurate and enduring alignment trio of all time. With regards to the laser collimator instructions, they gave just the opposite of the one step I later found it to be excellent for, especially when observing through fast optics.

'The Cave Collimator was excellent for my Maksutov–Newtonian and very good for my Rumak–Cassegrain,' said Professor Ian Morison, Jodrell Bank Observatory. 'It works!' said Peter Wise, on using one for confirming the Cheshire indications of the final alignments of his Zerochromat folded refractor mirrors and lens.

By using these instructions, a defocussed star that needs collimating never reaches the edge of the field of view without improvement being seen. The peephole, cross hairs, cave, and doughnut are the best in the early stages, all getting you to stage 3 with these same instructions. Then, so long as the primary mirror is slower than about f/4.4, the star is quite likely to show no need of further adjustment, or just a part turn of one primary mirror bolt. In the hands of beginners, both the laser and the Easy Tester respond unpredictably. A mirror or a bolt head is at risk of damage when you feel you are nearly there with these two aids because they can take you too close to the extremes of adjustment. It's time to stop for a cup of tea, to breathe deep, and to look again at the instructions to dissolve feelings of trepidation.

On now to fast primary mirrors after stage 4. These need three more stages, and the extra work is worth it because even Newtonian telescopes of f/4 to f/6 are better with these instructions than refractors faster than f/10, which cannot control false colour at the high magnifications needed for splitting close double stars. The problem is the *vertical tilt angle requirement* of the secondary mirror had become quite precise, but after three years of some success with black logic, I lost it. What to turn to now? The answer was to give a fourth and last chance to the laser collimator. Adjusting the beam to reflect from off the horizontal cross-sight line as imagined by the doughnut, by adjusting the vertical tilt bolt only, changed my perception of them from maverick to essential for this task. Rejoice, because back on a second-magnitude star magnified to high power, you are within two adjustments of concentric and sweet, all the way into focus. This now means the instructions are definitive. The confirmation was achieved during an evening of very good seeing and transparency on 2014 October 1–2, at half-moon, as follows:

STF3050 was not split, so vertical tilt was then set with the laser beam, creating the two final adjustments. Then ι Cass triple star was superb—no filters and no mask. With 72 Peg showing 0″.58 separation and a figure-of-eight shape, an estimate of the position angle was the same as in Cartes du Ciel and the British Astronomical Association Handbook. STF3050 was now a delightful orange pair with a black gap. It had never been that good since some diligent lucky collimating in 2009 ran out. Luis Argüelles's Difficulty Index in *Sky &Telescope* (Jan. 2002) is a good source of testing and of interesting double stars, as is the Webb Deep Sky Society. I would add ζ Cancri for a 4″. Then for the first time, M13 was more impressive than M92, and a week later the full moon + 1 day was really stunning with its small craters. Oh, for my imaging to do them justice.

The doughnut will show any offset between the optical and the geometric axis, and then there's the repeatability of tests. The twist-lock eyepiece helps a lot here and is much loved by haters of grooved eyepiece barrels. The actual instructions are in numbered military rote style so as not have you fishing amongst good bedtime reading when trying to launch the lifeboat. A hexagonal mask, by being able to correct for irregular little spikes masking close companions, can be a good indicator that a little more stellar collimation would be useful, and therefore you may do without the mask. Regular-looking spikes will be 'pinched' by the primary mirror bolts that need easing slightly off their springs.

A third source of complete support for my dispensing with two of the adjuster bolts was found only on 6 December 2017 on Google: 'The Ultimate Newtonian: The Secondary Mirror Holder Described and Explained" by R. F. Royce. Royce says the same in his own words.

CHAPTER 9

DON'T BOTTLE OUT!

Those who can't, predict. Those who can, do. Hands-on! Please don't live from the neck upwards if you do not have to.

Three aids for collimation appear to have peaked in 2014–15. Steve Ringwood in *Astronomy Now,* July 2014, was astonished to find, 'So accurate was the pre-stellar check with Howie Glatter equipment.' The May 2015 issue has the Catseye Infinity with a second peephole off-centre, so when a star test of a fast telescope indicates 'Use it,' both laser and doughnut reflections are given the illusion of looking aligned. And so by parallax and expense, the user is assuaged.

Being retired has provided me with the time over five years to discover remedies to the shortfalls in instructions for aligning the mirrors that also give tools somewhere to slot into when it is thought a tool may be useful. However, you still may not need anything beyond two translucent tools. One is a white 35 mm film canister in the focus tube, and the other, particularly when the light isn't right or there are no suitable spider spokes, is my invention fitted over the front, for which materials were becoming available 70 years ago except for rationing.

Nobody, including me, seems to have appreciated at first the correct start to stage 1. A plan drawing suggests the shame I felt for not noticing that one should begin by mechanically squaring the mirrors. They clearly need screwing out against their mounting plates, a process that can be conducted almost casually, then back again dead equally now on all three adjuster bolts to give scope for getting the tubes and mirror edges to show no overlap. That's all unless perfect symmetry just happens. And preferably the primary mirror edge mountings are all visible. Initially, I'd say two turns of the centre bolt inwards off the mounting plate or what you've discovered about the telescope. Then with slow primary mirrors it can be difficult in stage 4, the star test, to see anything still in need of adjustment. A very good sign!

Norton's new publisher's decision no longer to use anything by Arthur P. Norton meant relegation of two of the three adjusters of the secondary mirror no longer appeared in the later editions up to the final twentieth of 2004. The opportunity to reduce the number of bolts to be thought about and thereby to escape collimation by compensating for mistakes with too many adjuster opportunities should be taken, not mused over and then ignored. Those who wish to turn every bolt must be controlled!

The new approach, that is better, correct, or definitive instructions, means there is really no such thing now as a fast telescope being more difficult to correct, because contrary to slow ones, their aberrations at slight defocus can be seen easily, and five years has got us into knowing exactly what to do. Except for the lens and mirror making defects, problems of collimation are completely those of method. Tolerant slower telescopes have generated an appeasement column, which if dodged causes them to be misapplied to fast primary mirrors. *Collimating a Newtonian* provides all the instruction needed to bury all that. First for all reflector telescopes fully sufficient for slow ones, then without the 'maybes' and 'sometimes' of others, these instructions continue with just three additional steps, seven in all, to get fast and handier length-modulated telescopes working well for all objects of interest.

The star 72 Peg 0″.57 is on the Dawes's limit for splitting through an 8″, whilst STF3050, nearby, is a delightful orange pair for all and a test for 4″. *Try seeing what you can achieve with these close double stars before and then after having put your telescope through the instructions on a night in early autumn.* They are just 6° apart, straddling Andromeda/ Pegasus near to Alpheratz.

Eventually, by September 2014, the black art of setting the secondary mirror's vertical tilt angle of my 8″ f/6 Wise Newtonian had drifted beyond me. Despite its having been eclipsed for other uses over the years, my nine-year-old laser collimator rescued the situation well, contradicting its manufacturer. Anyone still holding out against some use of collimation tools in metal, laser beams, or serious plastic at last doesn't know what he is missing and by how much less he could be blaming the seeing and enjoying a reduced need of filters or masks for unwanted brightness or spikes. A curse on no tools, persistence, and luck for keeping us out of the dire straits for a while. Few people nowadays have the time.

Newtonian telescope collimation was swamped with too many words and diagrams, causing users to do it the wrong way! Double stars out of fashion can't have helped either. Then improved optics in the 1980s began to reveal collimation needs like the defects in depictions 5–8 and worse, with minimal corrective optics to halve the cost. At around 100 mm diameter, this can work well enough because there's little difference between parabolic and spherical primary mirror shapes. Then it's risk-taking or higher specification for 8″ diameter when corrector lenses are needed for making a handier length far more worthwhile than just for the Moon, only at low power.

Weak excuses for doing nothing. The only other feedback has been two cheques without comment from customers who had no cash on hand at events.

My observatory is not built yet. How much do you need one? With the slit-dome type, you will only ever see the fireball that has your name on it.

Three of us were on a lane between Ellerker and Broomfleet in East Yorkshire on the night of 2012 September 21 with our telescopes, when what should have come screaming horizontally out of the east across the northern horizon at about 5° elevation was a seemingly airborne express passenger train all lit up. Bright sparks, mainly green, were radiating from the 'engine'. It had a great long orange-yellow tail and must have been in view for about 20″, never expiring as the western horizon absorbed it. Our estimate put it overhead between the Tees and the Forth, and it was imaged over Consett. At a recent talk by a meteor expert, I was very sad to be the only person with fireball reports to offer. That was my lot too until 2003, when I retired from my aerial and satellite business.

If you can keep the telescope tube weight down to no more than 11 kg, and 14 kg for the rest, an HEQ5 heavy-duty mount and tripod, brought out for each session, can be less complicated. It can provide you with good exercise and will last for almost a lifetime. In addition, it's versatile for travelling by road.

It's too cold. Yes, warmer weather is far more preferable for doing a full collimation, and there are no gloves specific to astronomy, just dinghy sailor's gloves that have two fingers and the thumb missing from, labelled 'Astro'. So first my daughter bought me a pair of runners' gloves from the local T-Mac's. They have lightly reinforced thumb and index fingertips with mohair in them, so they're better than cycling gloves. Then I visited the boat chandler, wearing those gloves for oversizing, and bought deckhand gloves that have only one finger cut short. These gloves can be worn separately or together. In a previous year, I found the ones with flip-back finger and thumb ends to be a mixed blessing.

I've got to move first (because of street lights). Since you are now on a side with local street lighting departments, phone them. Speak to the manager and ask in a friendly manner how he's fixed for prioritising your request to reduce his costs with a change to flat, low-angle, moderate-power, yellowish LEDs. Also, with the screening of lights that interfere with sleep, the law is actually with you. The manager will be keen to know where the least resistance is likely to be. You do some useful work for astronomy. Every neighbour appreciates a bang on the door to switch a light off as well as an invitation to join you! I leafleted houses a quarter mile across a field where they don't have street lights, and was delighted with the response. Make a large enough black butterfly net and attach it to a 12- to 15-foot stick.

Support your local wind farm protest with signatures. The intention will be to put the wind farm too close to where you observe from! Then there's the angling down floodlights of light industrial premises. You don't wish to be enemies of nearby houses, or astronomers for that matter. Say a 500-watt bulb for ten hours a day over the course of a year costs £220 at 12p per kWh. Burglars need light and are helped by not needing to keep their activities to around the full moon. The security lights of the local large glasshouse premises are no longer visible to me.

It looks too difficult. You've read far too much beyond stage 1. Can you still only ride a tricycle? Dodge the column now, and you will be hampered by balls and chains and Achilles heels forever and a day. You will be arriving at the star test without first having kept the mirrors square, spouting about the difficulties increasing with the fineness of the optics and inversely as the focal ratio mumbo-jumbo, saying that with these instructions you get no real sense of these things at all. Hurray! Scrap your own cherished results so far with the twiddling of the three secondary mirror adjuster bolts and start again at stage 1, the best bolthole to run to!

The Moon at 36 hours after full. 1/50 sec ISO 200. DSLR on an 8″ Wise Newtonian through a Baader Hyperion 13 mm eyepiece.

CHAPTER 10

MAKING THE CAVE COLLIMATOR

Introduction

Just like every other tool for assisting with the alignment of telescope mirrors, none is the magic wand that can do it all for the Newtonian. Altogether, though, they each play their part. And despite what better optics now reveal, help with getting it all done before the star test is still made possible for classic slower ones. Not that you shouldn't check! The contributions of the cave collimator are that it has accurate cross sights and converts the wrong sort of lighting into uncluttered white clarity (see front cover). It helps indoor collimation become hassle-free so you may do it in comfort and style.

There are three ways of making a cave collimator. The easiest is to seal an end of the telescope, then apply glass-reinforced plastic, fibreglass. The mixing ratio is about a golf ball to a pea, and it is best to use a clear catalyst. You can begin with a first layer of tissue. If choosing to mould on to the front end, protect it with a plywood or cardboard disk, which you then cover with cling film; originally we used liquidised soap, which dries out to become the releasing agent. Wrap an average of three layers of insulating tape around the flange to 60 mm down, creating stepping for assisting the release and for a lessened chance of a hacksaw's scratching the finish.

You are wanting an inside depth of about 26 mm. This is the easiest method for getting the wetted material to hang vertically during curing, assisted by a retaining band of GRP, then masking tape. Finally, use a plastic trouser belt, a strip of plastic-edge veneer, or vinyl for a more even finish to the flange. This also means no need of felt-backed adhesive tape for the almost inevitable oversize of the other methods common to solar filters.

Alternatively, a disc can be cut from horrible MDF board to a radius of 2 mm to 3 mm more than flange radius, depending on thickness of felt tape, to be bought from specialist suppliers. Rounding the edge and a 1° taper helps with the release. To hang down sufficiently, it needs to be at least 50 mm deep, probably needing two pieces of material screwed together.

Initial cutting at least can be carried out carefully with a jigsaw, or you may prefer several straight cuts with a circular saw, then finish with a jigsaw and a good wood file. MDF board produces awful fine dust that goes right through collector bags like neutrinos. A block of oak will be far nicer to work with. Plywood may be good too. A neighbour will have a lathe to speed up the finishing. By any method, the fit has to feel firm but must be able to be rotated easily.

If this is to be you first go at fibreglassing, practise first on scrap such as an old plastic dish, or reinforce a pair of cheap plastic washing basket handles. Thanks to long-lasting motorcar underbody work since the 1980s, fibreglass is no longer the multipurpose material it once was. Thin latex protective gloves are essential, as are old clothes or a boiler suit. It's 95 per cent preparation to make the cave collimator. Allow two hours. During application of the fibreglass, do not answer the phone. Application can be done by dappling with an old brush or by using a small roller. The hand-roller method sets you exactly on track for making your own wind turbine blades, which are all coated by hand in glass fibre over the balsa wood shape.

A finished blank can be taken along to a plastics works for heat-forming the collimator from opal white acrylic sheet. Apollo Plastics have been in Hull since astronaut John Glenn's visit in 1964. They may have your size of blank ready. Sorry, no, this is not the farthest back in time the latest telescopes can make use of excellent old technology. The 798 1-metre hexagonal segments of the European ELT's mirror are to be kept in precise shape by the horse-drawn carriage's whippletree linkages. Lightweight they will be.

Tap in a hardboard nail head to mark the centre with a depression. There are various ways to determine the centre position, for example by using a spinning lathe and a pencil, or the point of a pair of compasses with the pencil end vertical to inscribe arcs from eight points around the inner edge. Also use a paper disc, for which a circle cutter is the best tool for helping with positioning the doughnut.

To mark the cross sights in with permanent ink, the obvious choice is a protractor. But these will be the wrong size and will slip about, so cleverly shape a dedicated ruler to fit exactly, made from something such as a length of plastic curtain rail. A protractor comes into use for setting the second sight line. The inside fitting disc is not the easiest DIY method it may appear to be in the photograph shown on the overleaf. Tissue applied first and last is optional.

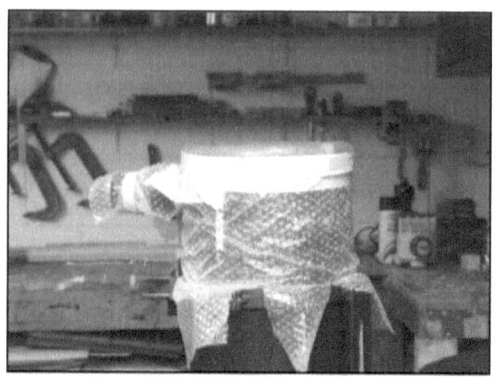

Tube end method.
Then cut carefully
with a large
hacksaw, angled.

Plug method after
removal.
Keep gloves on for
shards.

DIY. Colourless
catalyst if
possible.

Pressed-in works.

DIY cut from disk.

CHAPTER 11

SUGGESTIONS FOR MATCHING AND UPGRADING EYEPIECES

f/4.4 telescope				f/6 telescope
Eyepiece		**Exit pupil**		**Eyepiece**
30.6 mm		7 mm		42 mm
		wide field		
22.0 mm	Wide field for	4 mm	older person	24 mm
		Medium field darker		
9.8 mm		2 mm		12 mm
	Medium field + Barlow lenses for close double stars			
6.6 mm		1.5 mm		9 mm
	The eye works best for planets when placed 1.5 mm from the eyepiece			
4.4 mm		1.0 mm		6 mm orthoscopic
		For general collimating		
2.2 mm		0.5 mm		3 mm
	Crazy but can be needed for close or difficult double stars			

Medium-field eyepieces in between enable shorter focal lengths, which increase magnification. This darkens the background so fainter objects can be seen more clearly. Going by the Orion US 19 mm Edge-On Flat-Field eyepiece and the Antares 25 mm, a UHP Plossl, the optician has provided the field-stop diameter for 25 mm plus an expanded centre for 19 mm magnification. Orion's 16 mm, when used with a long ×2 Barlow, could be very good on planets and should be a better choice for f/5.

Use orthoscopics for high power for the least central distortion of planets and other bright targets and for collimating. Hence, true FOV is only 40°. The 1 mm to 1.5 mm exit pupil varies a bit with the seeing and eyesight. My preference just for planets is the Antares 7.5 mm Ultra Highly Polished Plossl.

Equations: True FOV = $\dfrac{\text{f s d} \times 57}{\text{Telescope } Fl}$ or   True FOV = $\dfrac{\text{Apparent FOV}}{\text{magnification}}$

where magnification = $\dfrac{Fl \text{ telescope}}{Fl \text{ eyepiece}}$

—— CHAPTER 12 ——

THE NEW APPROACH CONSOLIDATED

'You'll have trouble with that one,' said a telescope manufacturer as he spotted my Wise Newtonian whilst breezing past the stand at the International Astronomy Show in 2015 in Warwickshire. *Wrong.* It has more to do with trying to apply sensible but none scientific instructions that work well enough for slower telescopes to the better optics and faster primary mirrors from the 1980s that were failing, and so creating a market for tools that try to leave established methods undisturbed. Finding that my cave collimator invention of 2009 was not the complete answer either took me off euphoria and on to the task.

The more advanced the telescope, the more you must first check with the past to uncover anything that didn't matter then. So absorb the relevance of the horse-drawn carriage whippletree of the eighteenth century and apply it as the mirrors for both deepest and faintest professional use become bigger. Then, forwards to the 1950s, when *Norton's Star Atlas* relegated the secondary mirror's lateral adjusters with such polite none emphasis that it wasn't until after my independent realisation in 2012 that I had read the author's five-line paragraph and saw confirmation of what I'd realised upon awaking one morning. It is that independent fiddling with lateral adjusters acts like rotating or turning the handlebars of a bike if you don't lean over. Inertia puts you on the ground. So instead of remaining with the tricycle, enter stage 3, rotation again, but this time at high magnification on a star. Done. Then it was two years to 1 October 2014. The faster the primary mirror, the more secondary mirror vertical tilt adjustment only really proved to benefit from the laser collimator.

Of the first eight telescopes that came my way, two were unsuccessful at halving the length of the Newtonian, with one needing a boy runner's cleft stick to hold the nuts to get the bolts out to release the primary mirror for cleaning. The fineness needed for f/3 mirrors enables the slower ones to be fully collimated much more quickly. It took until the ninth one, a 10" Sky-Watcher f/4.8, for my comeuppance to occur.

It's owner drove to my house and I to his, where I phoned home and so then drove back to my house. This misunderstanding unwittingly caused the finger and eye trouble that prevented doughnut, cross wires, and peephole from stacking and our imagining that black art lucky collimating had advanced us two stages and, at the first star test, all would be well as usual. It wasn't all well. So fiddle, fiddle, it doesn't look far out, but the birdies, etc., were not to be cured by persistence, tools, or gadgets. So we scrapped everything, and after a cup tea it was telescope off the stand, back on to his garage bench, and starting again from the best bolthole to run to, stage 1 of *Collimating a Newtonian Scientifically*, after which the star test showed no further adjustments needed. The owner, our chairman, was delighted.

The answer to 'How do you collimate a Newtonian when the manufacturer has supplied only instructions for use?' begins by reference to a decent enough drawing, thus, (→ / ←) of the secondary mirror and the light path to the eye *V* produces the correct method for stage 1.

Clear to those whose maths need a half-decent drawing to be confident of their answers is that (1) the mirrors must begin from mechanically square relative to their mountings, and (2) there needs to be scope to adjust for the angle of reflection off the secondary mirror that achieves symmetrical star images. There is no profit in cutting or drilling twice. It's measure two or three times, set up the tools, then work accurately once, so that of any metalworking errors imagined, none will be found.

Here now are the logbook abstracts showing the facts and their key dates. The period **2008 December 7–8 to 2009 May 28–29** shows deformed star shapes during observing. There are moments of concentric appearance, but they are not maintained until 2009 June 2.

2009 May 17–18. Forgot to remove the three-year-old translucent fibreglass lid when about to collimate on the bench yet again. Eureka! Since 2006 January, I had struggled with the laser collimator's instructions, trying this and that, whilst not knowing that its manufacturer too was up against what may have been good enough for slower optics. Everyone has been conned into manipulating all three secondary mirror adjuster bolts just because they are there!

2009 May 28–29. Definitely preferring fine rotation of the secondary mirror to attempting the same effect by fiddling with the 120° lateral adjusters. Checking the focus tube alignment with an eye for what looks wrong, then to a set square, spirit level, and plumb line. If everything looks good, do not fix it! Please see photograph on page 10.

2010 March 26–28. First screwing of both mirrors out against their mounting plates to give a start on a base for no other purpose than to know where the mirror is. Nobody else mentions this for Newtonian telescopes.

2010 June 2–3. ε Bootes blue and yellow colour reversal corrected by a quarter turn of vertical secondary mirror tilt. Black art perfection at last, but needing four riveting extra stages for the f/3 primary of my Wise Newtonian in the first two editions. Reliability seldom lasting through to the next session.

2011 October 3–4. Brough Astronomy Society members buy a Sky-Watcher 8″ f/6 Newtonian for £200. A same-spec comparison at last! Separations are more distinct on the ι Cass triple star, but colour differences are seen only in the Wise.

2011 December 3–4. Sky-Watcher 8″. The B and C stars look of the same magnitude, but are better split, so it's natural that the slower primary mirror is both doing and not doing the business.

2012 March 1–2. Hitting on the importance of maintaining geometrically square throughout stage 1 by dead equal turns when screwing off the backplates. Seeing detail on Mars with the Wise Newtonian. Now better than Sky-Watcher's 8″.

2012 April 10–11. ζ Cancri's position angles showing correctly for the first time. Mars stunning.

2012 July 3. A very successful drive to Professor Ian Morison's house in Macclesfield, using his telescopes with cave collimators to hand-fit them.

2012 September 21–22. With two witnesses I tried to improve the collimation by following the laser collimator's instructions. Thankfully, I'd recorded the moves towards disaster their three lines got me to make, so then I could backtrack to finding the primary mirror at depiction 8 on page 21.

Between 2010 June and 2012 March, collimation of both telescopes was clearly within smug, easily satisfied quality. I am blaming the seeing for defects or avoiding them by keeping to faint objects, wide double stars, and the Moon. Blind to missing detail in globular clusters. Indeed, changing seeing can produce fleeting excellence, possibly aided by a hexagonal mask or filters, or even a thin cloud or a smaller telescope, so a bright and distorted star doesn't obliterate a fainter close one. The first edition is printed.

Settling for this state would have been easier, but on October 15, preparation for a rehearsal that night for a talk on collimation to the Kings Lynn Astronomy Society was weighing more heavily on my mind than young Birman cat training. Whilst sitting down for tea, our good neighbour knocked with bad news.

2013 October 25–26. This was the date when instead of fiddling with the lateral adjusters, I changed to *outrageous* *application of rotation at ×360*, using Enif whilst slightly defocussed. The difficult binary star μ Cygni was then split for first time: 4.8, 6.1, separation 1".55. 30–31st ε Aries also split.

2014 February 9–10. The colours of ι Cass are true and the same with both telescopes and seeing the magnitude differences. Dark Wise Newtonian telescope showing the B star too bright, so mask, filter, or smaller 6″ scope useful. It was fine though in Antoniadi scale 1–2 seeing as there is less scatter.

2014 February 21–22. A clandestine session with Brough's Sky-Watcher 8″ f/6. Its primary mirror retaining clips were all made visible, followed by surreptitiously applying out the scientific method whilst it was already beyond the well-defocused star-testing ability to show aberrations. Then through both telescopes, results on ζ Cancri showed both close A stars and B stars had become clear and distinct.

2014 February 26–27. Jupiter's moons well rounded and easily visible through the Wise and without spikes at last. SkyWatcher's 8″ f/6 was remarked on a few days later as never having been better for imaging the Moon!

2014 October 1–2. The black art of setting the vertical tilt of the secondary mirror had somehow left me, so *I gave a fourth chance to the laser collimator*, this time strictly in accordance with the drawing on page 25, which I decided was the correct way to go about it. After resultant small rotation and then one primary adjuster needing turning slightly, the results became superb on double star STF3050. Then to 72 Peg and the reward of a figure-of-eight split and M13 better than M92 for the first time. This adjustment was the first one of any sort needed for a year, so black art had given me a pretty good time really, with a little blaming of the jet stream noted in the logbook for September.

October 9–10. Moon full + 1 day, really stunning with tiny craters and with impact rays spreading everywhere. Except from my one-shot DSLR photography, I couldn't have wished for more. Collimation of the slower Newtonian hides behind a tolerance of up to 2 per cent before aberrations are seen to any sufficient extent. Instead of re-examination, I developed tools, when it would have been better first to establish the correct instructions without them. Only then look for help with a troublesome stage. In this regard, SkyWatcher's instructions for Cassegrain collimation reads the same for the first two sentences of stage 1. Then they diverge.

Dismissing independent adjustment of the lateral adjuster bolts of the secondary mirror as far-fetched is the same as trying to correct an EQ mount tracking error with more levelling. Also, there is the analogy of laying a bike over to turn so as not to fall off, which turning the handlebars would cause by gyroscopic inertia. With rotation around the centre bolt, results are in view throughout whilst adjusting the secondary mirror to the indications of a slightly defocussed, well-magnified star. Now it can be said that fast optics with the right instructions are easier to collimate because defects become more visible as you focus in, not less as with slow primary mirrors. Satisfaction at the well-defocussed star stage often depicted has become as fast as f/4.8, the optics having become more accurate, enabling handier length for the same performance.

Collimation as a subject of discussion isn't helped by the audience's being unable to see the results of what the speaker is telling the person at the eyepiece and spanner to do. And it will be overcast, keeping the sceptics spouting with their hands and money in their pockets and their feet up. It pays to postpone until a night when it is going to be worth changing the comfort zone to outdoors at the eyepiece!

Whenever the seeing is so bad that the star test cannot be carried out, the first three stages will have gotten your telescope to at least a very usable state for enjoying the conditions. Then plan the finalising for when at least the middle of the jet stream is somewhere else. If a fast telescope performing well should meet an off-centre doughnut Easy Tester or laser beam reflection and you are upset by this, it can be assuaged by parallax whilst viewing through a peephole eyepiece that is being moved across the face of the focus tube head. Buy one with a hole already drilled into the side for $145: money for old rope that works. Or remove the now offending doughnut and sell the laser! Quite a learning curve just when you don't want to be on one.

To summarise from the cave collimator realisation, the aberrations I encountered with an f/3 primary mirror Newtonian telescope could not be removed by following handed-down instructions that once upon a time worked well enough. The correct, better, and definitive way to align and position secondary mirrors was found from mostly within the box. The benefits make all focal ratios free in ten minutes of 'sometimes' and 'maybe' wanderings and really widen the range of seeing in which excellent performance can be achieved in a relaxed and seldom tweaking manner.

CHAPTER 13

IT JUST TAKES SOMEBODY NEW

Finally, by observing an exquisite ι Cass triple star on consolidation night, the one port laser collimator proved to be the only extra tool needed. The misuse of the secondary mirror's two lateral adjusters, which causes excessive credence to be given to collimation tools, has nothing to do with offsetting it towards or away from the focus tube by swinging or repositioning. Please see *Norton's Star Atlas* (11th edn, *c*.1950), p. 50, and 'Ed's Tracking Travelscope', *Sky & Telescope* Oct. 2010 page 70, for secondary mirrors with just one adjuster only for vertical tilt. Three or more bolts can be more supportive of heavier designs, and they seem to be universal.

On 1 October 2014, five years after the cave collimator realisation, my 8″ f/6 Wise Newtonian (Cape Newise) became excellent once the primary mirror, when set by a star, reflected the doughnut from 3 mm left of centre. The laser reflection then became 20 mm right, and the instructions for working with any collimator were completed. No adjustments have been needed since. Then the reflection of the cursor line inscribed on the peephole eyepiece was seen to be 10° offset from the true one!

By my applying + and – one-eighth sinful turns to the secondary mirror's two lateral adjusters, the cursors agreed. Then correction of rotation and primary vertical tilt on the bench got the laser and Easy Tester II to reflect from the centre without my ever having used them! Fantastic! The trouble was, back on Pollux the defects returned just as if having used a laser collimator according to its own instructions, which I find somewhat perverse, for example astroengineering instructions advising the use of a Cheshire eyepiece for the secondary mirror. But on that good October night, it was their laser collimator used according to my instructions that was extremely useful, the first time, for tilting the secondary mirror to reflect the beam off the horizontal sight-line axis, as shown on the drawing on page 25. Faster optics are bringing this tool into favour for vertical tilt only.

Returning the lateral adjusters to give the most concentric image possible using a slightly defocussed star at ×240 requires a left rotation followed by one primary adjuster to turn. What happened to the trusty doughnut? It went 3 mm to the left again, but this is sufficiently different because the cursors remained realigned. Easy Tester II and the laser collimator have followed the doughnut each in its own way. And that's how fast Newtonian primary mirrors, with their finer focussing, work through the 90° reflection angle. The important thing is to know that the extra collimation is not difficult. It's just that following the wrong order, leaving something out unwittingly, and then compensating can last for a night and lead you into false hope for the next clear night. Until the jet stream forecast became easily available, the seeing could be blamed too often.

Tweaking on attitude and temperature changes after a fast Newtonian has been put through eight adjuster alignments can be too regular, whereas expansion and contraction after five adjuster collimation is quite possibly more linear; nothing needs touching for months. Aperture and hexagonal masks and filters become needed less often, but they may still be useful for bridging jet stream intensity divisions and also help with too much brightness and focal ratio.

Lens Correction of Spherical Primary Mirrors

The rotational energy of the secondary mirrors needs nullifying during transport. You achieve this by placing the long edge of the telescope facing downwards and stowing it fore and aft, being more particular with the heavier ones. The mounting cradle should be of the dovetail-fitting type that can easily remain on the tube as an antiroll bar and with the tripod off to take up less space. To say that bolt fitting for a fixed telescope is all right for one that is to be moved around is nonsense because it takes too long. The clamping rings of the glass fronts benefit from being given three or four drops of Captain Tolley's Creeping Crack Cure to prevent rotation. Low-tack masking tape is good for dust removal.

What about a Classic All-Mirrors Newtonian?

On 2015 March 18 and 22, I applied the sin of lateral adjuster tweaking to my 6″ f/5 Newtonian. The seeing was 1–2 on both nights, with transparency at a magnitude of 3 and 4.5 respectively. The closing impression after the cloud cover interrupted the first night was of a half turn to each side being needed to see any distortion. Returning to mechanically square on March 22 in one-eighth-turn stages was needed, getting back to less than one-eighth off for distortion to vanish at f/5. Any need for further adjustment was beyond my perception. The distortion had been wide fan-shape horizontal, the sort that mechanically square prevents altogether. The paradox is that whereas the doughnut was most useful for imagining the horizontal sight line for tilting the laser reflection onto, five minutes later, when I was at 'I'm (smugly) relaxed about collimation and don't do double stars,' the position it took up was embarrassing for me, being new to correcting fast mirrors and reflections of the laser collimator and Easy Tester eyepiece. But peephole, Cheshire, and cave cross sights reflect concentric to the millimetre, as in the drawing on page 25. For the slower ones that this paragraph started off being about, heaven is in Figure 3 and the images on page 12 after star testing. Both laser collimators can just be seen reflecting from a slightly offset position towards the focus tube.

It is reassuring that slower primary mirror telescopes can be fully aligned after accomplishing stage 3, so a slightly defocussed star showing a need to turn one primary mirror slightly should be no bother at all; see page 21 (depiction 8). When about to screw the secondary mirror off its mounting plate again for the first time, the second step of stage 1, you must suddenly become changed into the finest bolt manipulator going. A cup of tea is all you'll need to help start achieving the two, as found or advised, number of turns off the mounting plate exactly. Two identical telescopes may not require the same. Enjoy!

Neglected mechanical squareness is 25 per cent of collimating. It is a regular instruction with Cassegrains and does 50 per cent. Nobody writes of its importance for the Newtonian. Eventually it occurred to me in 2012, after a casual go in 2010. Jupiter and the Moon kept me off ζ

Cancri, but ι Cass was superb. And 57 Cancri first showed good separation, then clean definition with a hexagonal mask slipped down the dew cover. If secondary mirror vertical tilt is a problem, the simplest corrected laser collimator available, when viewed through the front *after* switching it on first, will set the mirror to reflect the red dot off its horizontal cross wire.

What about the Catseye Infinity XLKP Autocollimator?

A white translucent film canister as a peephole eyepiece is easily seen to respond immediately to the slightest turn of a primary mirror adjuster, but my Easy Tester II autocollimator needed a three-quarter turn before complete blackness started to become a partial eclipse. The site at www.stargazers.net, under the thread heading 'The BEST Autocollimator', has the Infinity XLKP showing circles and triangles improving on the clarity of black plastic or metal. For no worries, though, always finalise with a slightly defocussed and highly magnified star. The second peephole of the XLKP appears to be at an offset position, the purpose of which can be simulated by passing a normal peephole eyepiece across the top of the focus tube. Its slanting view has stacked the slight disarray centrally by parallax, and so the tormented are assuaged into agreeing when a sweet and symmetrical slightly defocussed star screams at them, 'Touch nothing!' they will set the mirror to reflect the beam off the horizontal sight line as shown on page 25.

Now is not the time to respond with a spanner to the reflection position of a laser beam, a cave collimator, or any other tool. With central stacking of the doughnut, the laser collimator may be harmless up to the nearly 2 per cent tolerance of slower optics. For a fast Newtonian at stage [4] and showing aberrations, stages 5–7 are there to help you complete gracefully.

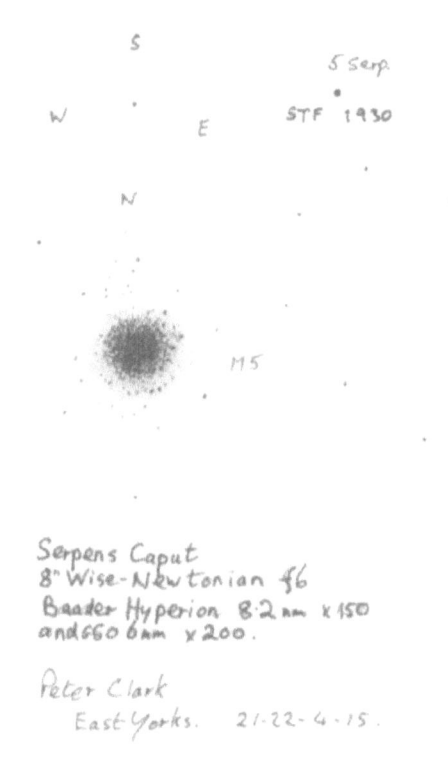

8″ Wise Newtonian f/6 with a Baader Hyperion 8.2 mm ×150 and a GSO 6 mm ×200. East Yorks. Limiting magnitude 5.3, seeing 2.

ACKNOWLEDGEMENTS

I would like to thank all the following people and organisations for their contributions: The Boy Scouts Association for the Astronomy Badge. Then my mother, for buying me a *Norton's Star Atlas* in 1954 for 17/6p. Norman Hully of Bradford University in 1975 for saying to me, 'Never mind if you don't understand the equipment. Use it. Understanding will follow.' Thanks to my partner Kathleen for spotting a bungalow for us to retire to, prompting the buying of my first telescope. The plastic lid cracked, causing me to make a new one quickly in fibreglass, which three years later suddenly made collimation on the bench easier. Martin Hornsby for pointing out there are persons like him for whom blood flows whenever they touch cutting tools. He gave me the message that I make neat Fuji film canister peephole eyepieces for sale. Professor Ian Morison for the cave collimator results on two of his telescopes, and subsequent conversations. Julian Robinson for his website efforts and for using his lathe on my behalf, and later Steve next door for lending me his. Apollo Plastics, Hull, for their professional touches. The work of Arthur P. Norton to 1954 and Ed Jones in 2010, the articles of both of whom I knew nothing of until 2012, when I was desperate for support for my drubbing of common sense collimation. Brough Astronomy Society for requesting £20 each for buying a Newtonian with identical specifications to the mirrors and lenses one most of my collimation developments has been made with. Tony Scaife for subjecting his 10" Newtonian to the fourth edition very successfully and for some lines in the handout pamphlet suggested by the organisers of the International Astronomy Show. East Riding Astronomers, Beverley, for putting up with my talks, and for buying and using. Peter Wise for keeping me optically straight and for buying a cave collimator for the checking of Cheshire eyepiece indications through his Zerochromat folded refractors. The organisers of events for granting me space to set up my stall. Last but not least, John Watson, Springer's UK authors' agent, for written advice, even though the subject cannot now, thanks to the new approach, produce anywhere near enough pages for Springer to become my publisher. John has actually given me a complete list of headings for such a book! That's ten years at my more prolific writer speed.

APPENDIX A

TWO MEDIUM-FIELD EYEPIECES

Summer is an excellent time for checking things out because the limited darkness encourages decisions. Skies often clear late, and the seeing quality frequently becomes Antoniadi scale 2. And if the relative humidity is 85 per cent or less, it becomes well worth seeking a dark site with clear view low to the south.

Dissatisfaction with low-power fields of view for locating objects of interest and medium-field viewing caused me to look for new ones. Some less than top dollar priced wider-angle eyepieces with better compromises had become available, and high inflation seemed about to strike, as it was high summer, whilst the East Riding Astronomers' eyepieces were available to check purchases against without waiting for reviews. Limiting magnitude 4 became 5.1 or 5.2. There was no moon, but up to half-moon at this time of the year makes little difference in clear skies away from the direct glare of the Moon. The need may occur when it has been decided the price of eyepieces supplied with a new telescope has to keep the total price below a certain level. So a lot of buyers will get what have become value-for-money interesting museum pieces. Indeed, the *Norton's Star Atlas* of 1950 describes as very large the 40° field of view of Kelner eyepieces!

The first task was to number-crunch the specifications into showing the true fields of view through both telescopes and then note the claims, weights, and prices of the eyepieces. The best equation to use is $FOV = \frac{FSD \times 57.3}{Focal\ Length}$ because field-stop diameter, the clear measurement across the back of an eyepiece, is easy to get from stockists, and it gave conservative results with my 8″ Wise Newtonian. Here is what I came up with:

My first choice was an Antares UPL 25 mm seven-element 52° Plossl for £70, because (*a*) it would reduce changing between the red eye finder and the 8 × 50 finderscope because of only an 0.8° field of a Celestron Zoom eyepiece at 24 mm through my 8″ f/6 reflector; (*b*) the zoom eyepiece was rather too heavy for the 4.5″ Bresser 500 mm f/4.4 Newtonian, and did not give as much low-power wide-angle viewing as a short 'scope can; and (*c*), it would provide (by Barlow lensing) double star fields for when referring to *A Visual Atlas of Double Stars*, by Mike Ropelinski. This eleventh-hour penny drop clinched a difficult choice online from a short list of 13. The well-reviewed Meade 26 mm had more field of view than I needed. The Antares UPL has never beaten at what it does in a manufacturer's test.

My second choice was the new Orion (US) 19 mm Flat-Field eyepiece costing £90, for medium wide-field use with the 8″ Newtonian—'optically aimed at providing a pinpoint focus across a flat field'.

The aims have been fully achieved for the purposes of each one. Here's what I saw when centred on Rasalhague:

Full field in focus at f/4.4	Achieved only with the Antares 25. Orion 19 impressive for the first 60 per cent.
Full field in focus at f/6	Both give good focus at the reduced apparent field of view in the longer focal length.
Faintest stars seen at f/4.4	Orion Flat-Field 19 mm, the slightly greater darkness by paying off in the similar actual fields of view.
Faintest stars seen at f/6	Both about the same without seeking anything fainter than seen with the f/4.4.
Focus sharpness	The Antares was sharper through both instruments. Orion 19 was not so plashy that it could fail to impress for general viewing. It was a little easier to focus and was first to reveal the Jovian moon Io close to Jupiter.
Eye relief and comfort	Orion, but the slip-on eye shields of the Antares work well. The Orion twist-ups are nice to use and better than the Teleview Radian twist-ups until you know your clicks. The Radian is my only expensive eyepiece.

The brief early July viewing concluded with my locating M12 at 30° altitude quickly in both instruments, then Barlowing them out on the Wise Newtonian for comparison with the Radian. Never mind the differences, I felt that in the time available, and given the changing darkness of the time of year, any of the three would have produced an image like the one taken through a 10″ as shown in *Atlas of the Universe*. The 10 mm Radian eyepiece is my close double star workhorse. The other is a Meade 12 mm illuminated reticle.

On 2008 July 9, I awoke up at 0150 BST to the clear skies forecast to be high cirrus. The observed FOVs by timing a star within 10° of the Celestial Equator across the eyepieces with the Bresser were 2.49° Antares, 2.51° Orion, indicating just over 1° with the Wise Newtonian.

For 2″ wide-field eyepieces, the ERA's 32 mm Orion Optilux enabled me to say 'ugh' to a £130 Antares modified Erfle eyepiece for aboard-the-starship viewing. I found its 1960s design no better at anything than the £50 GSO 30 mm supplied with the telescope, so I exchanged it for a delightful Moonfish 30 mm 80°, giving a true 2°, for £82.

A year later, these two eyepieces are the next to be used after aiming with a red eye finder, resulting in much less need of the excellent but cumbersome 2″ Moonfish, or changing the red eye for a 5° right-angle upright finderscope. Then they are put away until one set of demanding, faint, neglected double star measurements has been accomplished; experience and 70 years has not let me get confused by trying to measure two in one night. Into more relaxed viewing, they come out again the more my interest goes onto wide- and medium-field viewing objects, the Orion eyepiece when central detail is more important, the Antares for everything else.

So I'd say that for any observing and imaging society, having a range of library eyepieces to borrow and compare with a purchase is a very useful benefit of being a member when you can't find any reviews on the internet or in magazines. Or if you can, they are too kind, are played down, or simply are not judged through your telescope by you. Society eyepieces don't have to be the latest, just good enough that price and requirements can be judged and help members make decisions to keep or return theirs and then go on from there with the retailers.

It is most reassuring to note that several manufacturers have responded to, 'Hey, this telescope is worth rather better eyepieces than the ones supplied,' without pitching with interest specific ones. Had I just one telescope to satisfy, the Meade 5000 20 mm Superwide could have been tempting at £180. For slower optics, there's Orion's 27 mm Flat-Field eyepiece and Baader's Hyperion 24 mm, to which you can attach your camera's body, giving you a stout telescope. I find their 13 mm gives attractive views of star clusters, and cheap ring extenders are available to increase magnification without adding glass.

APPENDIX B

A HANDY CATADIOPTRIC NEWTONIAN

It was in December 2004 when I ended 50 years of astronomy by being steered by *Norton's Star Atlas*'s basic information section into the purchase of a Bresser Pluto 4.5″ f/4.4 reflector, to test my interest. The exceptions, of course, were seeing the Halley and the Hale–Bopp comets. Also I had a 'this is the end' feeling when on a roof during the last solar eclipse whilst doing some TV aerial business, the accidental coming together of my past as a master mariner, exporter, cottage industry loudspeaker maker, and mountaineer.

Brick driveway observing is the only downside of my home site, ten miles west of Hull, but the sun is off it three hours before sunset, and my workshop storage is in an adapted sink unit six paces away. Limiting magnitude can be to 5.7.

By December 2005 I could not resist Peter Wise's mirroring of my own style of advertising a loudspeaker design in the 1970s, so I bought a Wise Newtonian 8″ reflector, as most people now seem to want him to call it. Despite a focal length of 1200 mm, the design doesn't need steps, nor two people for collimating.

The main trouble with the optically flat front glass suggests I've learned more about telescopes free of charge,

Peter getting me particularly interested in close binaries after star collimating. He has been awarded the Horace Dall Medal by the BAA for showing a marked ability in the making of astronomical instruments.

Once through the testing capability hoops ex-works, you get sharp star images, says a friend with a 12″ Dobsonian, which has a darker background and minimal central obstruction, is easily transported on the back seat of a car, and is optically flat over 1°. I get 6 per cent to 15 per cent more than formulae with eyepieces offering over 52° apparent. The optics produce a more parallel beam onto the secondary mirror, as well as have a shorter tube for the focal length.

Then at the focussing tube entry, the light is focussed into a flat and really aberration-free image plane. Front glass mounting is more likely to need a dew shield and heater, but then if there's a 12V output for the equatorial motor drive with the voltage converter, you'll save on batteries. A humidity indicator helps too, such as with the car window's degree of dewing up, as does a cheap hygrometer. In humidity of about 80 per cent, the cardboard shield alone can resist condensation for up to three hours, more in the light breeze of an open site.

The front glass problem is that it is easier to make curved glass. Usually, the cutting off the block of raw material with one saw blade produces two parallel sides. Two parallel blades are more reliable, but it's expensive and very wasteful. A millionth of an inch out in the polishing and distortions are seen in star testing.

All basic methods of collimating seem to have variations and omissions according to the problems the writers did or did not have. So see how far you can get with probably your second telescope, pretending you are Isaac Newton learning for the first time.

Upon weakening, I was recommended to try the laser collimator. This tool seemed to have been a good choice because after a bumpy cargo plane flight through turbulence from the United States, the device itself had to be collimated. My mounting method for aligning the beam was to place the device on top of a vice. The vice grips a piece of wood, the width of which is such that the switch knob can be rotated 120° each way from vertical (see illustration on p. 39). One with a flange will require a vice top a saw can cut a gap through. Check that each adjusting screw is reasonably tight, then project the beam onto white-faced cardboard about six feet away (two pieces come with a new Planisphere), marking its position when the knob is rotated left, right, and upwards. As you get closer, do a drawing to estimate the direction and the amounts by which to adjust just two screws from inspection of the beam hits, and end up after several repeats with the red spot not moving after a complete rotation. You have learned how to collimate almost blindfolded because there was nothing useful to look at whilst making the fine adjustments. Mine benefited from reducing the beam width with a section of model shop brass fuel pipe pushed into the exit hole—all good stuff. After seven years, one good use was found for it.

Then a new tool for collimating just happened to be invented by my having had to make a new cap for the telescope in fibreglass. You can mark on it cross sight lines, and it is used for bench collimating.

There were two 20-minute trips and one 50-minute trip to the darker sites of East Yorkshire in 2007. One was for tests on close double stars.

Results: Sigma Tri. Sep'n 0.6" was split clearly with a 13 mm Baader Hyperion and ×2 Barlow lens and with distinct colours on 2007 February 7, in 5.8 limiting magnitude; this was elongated only in October to 5.6, before I'd bought a 10 mm Antares UPL eyepiece. This did a superior job with a ×2 then a ×3 Barlow on ι Cass during a testing of four eyepieces. With the Celestron Zoom at 10 × 2, the provided 6 mm GSO Plossl was not far behind. The 15 mm Plossl ×3, the hitherto workhorse, was fourth, but please allow for the odiousness of testing at different times. The south polar ice cap and two-tone colouring of Mars were visible two out of four diversions off double stars. I keep telling myself I really didn't split S. Tri., but they were tiny, close, and about equal. Definitely not 6 (ι) Tri. The zoom eyepiece's field of view has given insufficient real field of view at its lowest power. For example, I can't see all the Pleiades in either instrument, but with a 25 mm Antares UPL Plossl for the Bresser, and an Orion 19 mm Edge-On Flat-Field 65-degree eyepiece, I can. I gathered information on the internet and over the phone, then did the calculations for 13 eyepieces, before choosing produced the right ones. A 10 mm Radian eyepiece is used for double stars.

Selections were also made with reference to 'Choosing Eyepieces', by Al Nagler, which came with the Radian.

Avoided so far is 'glass fronts versus minimal stretched metal spokes support for the secondary". Glass mounting prevented a clinker from the power station 20 miles away in the bad old days, landing on the primary mirror, and for this shorter than usual Newtonian it was more important than the disadvantage of pressure-held corrector glass plate. It can rotate. For me it took a year to happen, and seemingly just after two careful trips six miles away. Captain Tolley's Creeping Crack Cure around the perimeter in three or four places is the answer. Ease of rotating the main tube is a big advantage, so I hand-sanded half of the nice hammered finish and then repolished it to almost car body standard. McLube is needed annually—worth trying first.

So, I think we can say that pioneering Peter Wise really deserved the accolades he received. However, his 200 mm reflector did look suitable only for professionals and amateurs of his ilk, until the stumble on to the new collimating tool, which works the same way as the Cheshire and the sight tube, reintroduced by Ron Arbour in the 1980s, and is much clearer. Stow a Newtonian vertically, but travel with the focus tube upright and facing forward, otherwise *any* Newtonian secondary mirror will always want to rotate into equilibrium, and in time it will succeed. But that won't part me from what really is the excellent instrument described in the three UK reviews. Not in the United States during the years of wrong instructions for secondary mirror collimation, mainly because cargo planes dodge only violent turbulence. First-hand experience of this precaution joyfully occurred as one-third of a cargo plane containing the ships' officers group I was part of was offered several double whiskies before leaping and diving through very heavy turbulence whilst over Burma. You can imagine the suffering of any telescopes on board!

A laser collimator beam streaking through the central hole of it's diagonal mirror, or by achieving magic blackout with an Easy Tester 11 with fast optics, only means you are getting there. Well done! Likewise, computer programme cross wires that Chapter 4 stages [3] or [4] achieve without having used either, will show as unlikely by a slightly off-focus star at high magnification, to be final collimation of a fast Newtonian. The given tool for correcting small errors of any sort at final stage [7] is a star, when the Jetstream centre is somewhere else.

Peter Wise's customers were not helped by the editors and contributors to *Norton's Star Atlas* after the 15th edition of 1966 polluting ADJUSTMENTS OF THE NEWTONIAN REFLECTOR by replace his page 50 with their intuition and horror drawings of mirrors in anguish users of the scientific method never see. Indeed, recent star atlas/reference handbook author Robin Scagell, and Ade Ashford in an Astronomy Now article both pass collimation to Gary Seronik on the internet. He also spoils his very convincing website by ending wrongly on the secondary mirror; and he was ghost write for one bolt *Ed Jones's Tracking Travelscope, S & T* Oct. 2010 page 70.

Well, it took me two years of wandering to suss out the importance of the maintaining what my stage one achieves.

If you make the two lateral adjusters strict followers of the central bolt you bring the number of adjusters down from 7 to 5 over both mirrors. At 7 the human brain is unlikely to make the correct ones first time. On their websites, the three American supporters have stopped mistakes by redesigning the metalwork. Sceptics have never done any fine metal or wood engineering.

With the makings of the scientific method being all there by about 1950 this book should not have been needed, but....! And from even further back in time, refinements of the *Whippletree* balancing interconnections for multiple horse drawn carriage linkages are being fitted to the 900+mirror segments of the E-ELT for controlling atmospherics and sag, glass only being nearly a perfect solid. Marketing speak, adaptive optics.

More Like Minds Found from 2017 Who Also Corroborate the Scientific Method of Collimating the Secondary Mirror

R. F. 'Bob' Royce www. His ultimate Newtonian, *c*.2000, replaces the two *incorrect and illogical* lateral bolts with a single adjuster.

Conrad Hoffman. His 1990s secondary mirror bolts 'the dastardly pair' into strictly following the central bolt.

Dion Heap. *Advanced Newtonian Collimation*, minutes 8 to 11, on YouTube, February 2012, is identical to my first edition of March 1st 2012.

Rendering Newtonian Collimation Free of Fiddle Fiddle!

Vic Menard, in *New Perspectives on Newtonian Collimation* (5th edition, 2008), advises on page 14, 'Keep off precise symmetry of optical and mechanical elements,' but his other instructions are not clear to me.

I would say that with common sense and intuition not appropriate to the 45° mirror, the scientific method had to be discovered. It is identical to the drift method of correcting EQ mount tracking by rotation in azimuth. And also the cornering technique when riding a bike by laying it over, not trying to kill inertia by turning the handlebars. The handlebars only follow the turn.

A three-legged stool, a spirit level, and a thin shim can be used to simulate aligning the secondary mirror, smoothly by rotation, jerkily overshooting by lifting a leg. Advice to turn both adjusters in opposite directions can be wrong because only one may be out!

So, for Newtonian secondary mirrors, forget your mastery of primary mirror collimation, the Cassegrain secondary mirror, the laser collimator, and the universe. However, some of the multitude of wrong instructions can be very good on preparation.

APPENDIX C

CELESTIAL TRUANT CATCHING

Three years into observing with a telescope, I wished to do something useful in astronomy without spending a lot of time processing images, so in December 2007, I helped to determine the fate of the universe by checking faint, neglected double stars, which became my contributing interest. One comment from behind the BAA stand at Astrofest was a friendly 'Good for you.'

When I am seeking brighter or faint big objects, all else nearby goes unnoticed, whereas when I am hunting for a faint small target, my eyes wander more and sometimes can pick up remarkable new asterisms.

Finder equipment when a red-eye hasn't been sufficient, often at this faintness, includes a pair of 10 × 50 binoculars, a 114 mm Newtonian or a 2" monocle finder 5° at f/6, and a 30 mm 80° Moonfish eyepiece giving me 2° true FOV exactly. With the almost parallel presentation through the Wise Newtonian spherical primary mirror corrector lens onto the secondary mirror, I find no need of a coma corrector. Once the target position has been identified, I'll change to an Orion 19 mm 68° for 1° FOV, start the motor-driven EQ5 mount, and change quickly to a 12 mm Meade reticle eyepiece to measure separations. In calibrating one for the telescope, having done the honourable thing of putting the reticle eyepiece through 'Each scale division in arc = drift time in seconds × 15.041 cos. dec. 10 times + doubtful ones for each Barlow lens: nine times at ×5′. I confess that after six months I was asking the Webb Society for the latest annual check separations.

You choose your targets from the Washington List of Neglected Double Stars by asking the administrator to email your choices in magnitude, minimum separation, hour angle, and declination limits. Sometimes plotting onto the Cambridge, the Webb, or Norton's atlas is sufficient. Printing Patrick Chevalley's *Cartes du Ciel* off the internet is the certain way forward once you are hooked. Laptops may not agree with what some dealers say about longevity in the damp and cold. It's best to take them back into the house, likewise hand controllers.

A typical good example is **ALD121** of 2013 November 9 and 25 situated at 2352+5133 in Cassiopeia. The area was most easily entered with guide star 18 Andr., magnitude 5.3, red-eyed in a triangle formed with γ and 7 Andromeda. I seldom star-hop. Then place 18 Andr. at four o'clock, towards the edge of the 5° finder, look 1° N and 2° W to the centre, and you are there! Notwithstanding that declination was vertical and inverted by being about six hours west of the meridian, the first striking surprise was a vertical St Andrew's cross asterism that rotation caused to vanish as regards shape. Then the striking STT251 appeared as a bonus to the diligence needed, and then a third, unrelated star, making it look triple. With these two distinct objects in the 1° field, the target was surely going to be no problem to find. *It wasn't there.* I felt the need to tie on to something. About 7′E there was a pair with a magnitude of 12, so I settled on the idea that the suspects had changed thus in the 97-year interval since discovery in 1916 and measured them as best I could by inspection with a Radian 10 mm at ×200 as they were rather too faint for my reticle eyepiece for separation figures.

In the field, there was a feast of three other faint doubles. ALD123 I recorded in position, finding all similar to 1916. I observed ALD7 as a single star in position and of the right magnitude and ES2735 as intact. Relax now and take a look at the Blue Snowball, then other easier, impressive targets. Really deep-sky observers will enjoy Abel 84 glowing at magnitude 14.4 in the 2° field at PA 240 or ESE.

So what does the 3.8 version of Cartes du Ciel show in the ALD121 area? It shows that in 2000, somebody placed it intact with slight measurement differences over two stars off the nearest corner of St Andrew's cross. He'd also moved ALD123 to a pair existing alongside STT251, whereas I found it still in its 1916 position. Even more interest in a familiar field for ensembles of dedicated enthusiasts!

Now for **HJ943** of 1820 and 1909 in Pegasus at 2147+2648, 2010 August 31, and noting its pedigree of Sir John Herschel and his descendant Colonel John Herschel. The chart showed the way in via an attractive showpiece comprising the yellow κ Andromeda with fainter blue and orange stars in close linear company. From kappa, three celestial street lights of magnitude 9 aligned NNE, suggesting star-hopping. The location was going to be in the centre of a small rectangle of four magnitude 10–11 stars slightly off to the east of the third lamp post. Easy. Alas, the report was of an empty and giddy void in the galaxy. Time to show that I can avoid treading on other people's toes! It was also reported by Richard

Miles and in Hungary by Erno Berko. I offered nautical astronomy's clerical errors. They had caused one or two German villages to be bombed even after my father had instructed celestial navigation courses at Staverton, Gloucestershire, England, using Anson trainers, bubble sextant, chronometer, pencil, and rubber, which could be accurate to a square or triangular nautical mile five minutes after shooting the final star. Two of us once made different errors that produced the same wrong latitude at noon, then Trinidad appeared an hour sooner than expected in a run of an hour at 15 knots and 10 nautical miles to the west by radar. Wise were the shipping companies before radar that always had four officers observing at noon.

Checking for errors in declination of 1′, 10′, or 1° and in RA 6′, 10′, or 1hr drew all blanks for me. If it can happen, it will, and four years later in the case of HJ943, where Cartes du Ciel 3.8 version had the answers. Others had traced the magnitude-10 pair to 35′S, and ALD121 above at 60″ westward. Of course, there's a time lag between reports received in Washington, DC, the changes being published and amateurs updating the data. What will checks over time show? Do we have a celestial tramping binary pair? Either way, helping to establish the fate of the universe shone brightly through until the Sloane Sky Survey got up to speed, reducing the number of neglected doubles on which Washington seeks information from amateurs to mostly the ones not detected by the survey. So now may remain only fascination for Sherlock Holmes minds in the manner of the clues here described. To see deep clearly enough, 8″ at f/6 is about the right specification needed, true fields of view for finding objects being easy whole numbers.

The intensity of faint neglected hunting is almost guaranteed to be an automatic planner for 'What's next?' in the wider area around which you've just found nothing at all! Via faint neglected double stars, favourites will multiply.

APPENDIX D

INTUITIVE COLLIMATION DE-BUNKED

'You've made my day,' replied American telescope maker Bob Royce to my email sent after discovering him by googling for more published support than two in the world.

Steve Ringwood has commented on the parlous state of many of the instructions for aligning telescope mirrors and Ian Morison has agreed with me his last effort in AN was not all good.

To collimate a Newtonian secondary mirror, also known as 'the flat,' it has been intuitive to turn any of the usual 3 mounting bolts in order so to do. But owing to the 90° direction change to the light path off the primary mirror, instructions should not get you doing this universally popular common sense looking approach. It is completely wrong and actually more difficult. You get sent sent towards maybes, sometimes and occasionally hoops in the hope getting there.

By the TLK of *f*/5 and slower some do, but by analogy if you know that levelling beyond spirit level results is not prescribed for tracking errors of satellite dishes nor equatorial mounted telescopes. Try riding a bike around a corner solely by turning the handle bars and you fall off (the collimation). It is so unscientific that some amateurs fit only the adjuster bolt for vertical tilt. Both for self respect and commercial metal working, machining accuracy is sufficient to put lateral adjusters beyond usefulness, even for close double star enthusiasts like myself.

So upheaval and feather ruffling was long overdue. Rotation sorts lateral zeal gyroscopically because only one bolt has needed loosening slightly. At least two are eased for push pull suck it and see ventures, praying all the time for no unwanted movement upon gingerly retightening or hiding in low magnification.

Vic Menard's page 14 of his, *New Perspectives on Newtonian Collimation* blessedly keeps us off precise positioning of optical and mechanical parts. No overlaps are what you first need to achieve, I think is what he means by, *It doesn't constrain any of the optical or mechanical parts to a precise* position *or perpendicularity.*' I put it like this. 'As soon as the mirror and tube perimeters are no longer seen to overlap and the primary mirror clips are all visible, go to stage [2]. Mirror angles for achieving off-set without moving the secondary just happen. Or go with offset = Secondary mirror minor axis diameter / 4 x focal ratio.

I too had become a priest of dismissing the laser collimator until October 2014. *The other two adjusters are needed?* Rubbish. Like columnists caressing lateral adjustment and still saying one should attempt vertical tilt blind as a bat, praying and over correcting more as *f* ratio reduces. So to my laser collimator a 4th chance in 8 years was given. For the one purpose of visibly setting the angle of its reflection to off the horizontal cross wire as imagined by the central doughnut, it proved to be the perfect way to achieve this. Lateral or shunting adjustments are now avoided with or without metalwork. Simple, quick, accurate and reliable, the *dastardly thing* is tamed.

I am most grateful for not having the sort of mind that would have blamed the optics of my main development telescope, a Peter Wise Catadioptric Newtonian, and being retired having the time and towing my sailboat past his works in North Wales. Beginners and slower telescope users can be into use after completing stages 1] to [4] in 10 minutes, using a translucent 35mm film canister eyepiece from some of the remaining photographic shops. Those from Amazon have been too wide. The more experienced will rejoice at finer stages [5] to [7] for having no wriggling nor anguish over fast optics, to give excellent results at high magnification. Clear skies to get you quickly to these final stages and so into full use and enjoyment.

BEGINNER GUIDES TO OBSERVING

Any booklet or magazine containing monthly star charts.

Kraul, Walter, *Astronomy for Young and Old* (Floris Books).

The real-time charts at www.stellarium.org are free to download.

Bridging the Two

The Cambridge Star Atlas (2011).

Norton's Star Atlas (1910 to 2004). Available second-hand. Excellent reference sections that did well at being for all interests in one volume. Charts in nine slots of sky to magnitude 6, easy for beginners, intended for students.

Philip's *Planisphere*. For every night of the year.

Intermediate

Cambridge Photographic Star Atlas.

Cartes du Ciel, Patrick Chevalley's star charts for all and special interests to magnitude 16.5 if you want to. Free for downloading.

Cambridge Double Star Atlas (2009 and 2015).

Dunlop, Storm, *Atlas of the Night Sky* (Collins, 2005). Detail by constellation.

Morison, Ian, *An Amateurs' Guide to Observing and Imaging the Heavens* (Cambridge University Press, 2014).